OBD II: FUNCTIONS, MONITORS, & DIAGNOSTIC TECHNIQUES

Al Santini

DELMAR
CENGAGE Learning™

Australia • Brazil • Japan • Korea • Mexico • Singapore • Spain • United Kingdom • United States

DELMAR
CENGAGE Learning™

OBD II: Functions, Monitors, & Diagnostic Techniques

Al Santini

Vice President, Career and Professional Editorial: Dave Garza

Director of Learning Solutions: Sandy Clark

Executive Editor: David Boelio

Managing Editor: Larry Main

Senior Product Manager: Matthew Thouin

Editorial Assistant: Jillian Borden

Vice President, Career and Professional Marketing: Jennifer Baker

Executive Marketing Manager: Deborah S. Yarnell

Associate Marketing Manager: Mark Pierro

Production Director: Wendy Troeger

Production Manager: Mark Bernard

Senior Art Director: Benj Gleeksman

Content Project Manager: Pre-Press PMG

For product information and technology assistance, contact us at
Professional Group Cengage Learning Customer & Sales Support, 1-800-354-9706

For permission to use material from this text or product, submit all requests online at **cengage.com/permissions**
Further permissions questions can be emailed to
permissionrequest@cengage.com

Library of Congress Control Number: 2010925169

ISBN-13: 978-1-428-39000-3

ISBN-10: 1-428-39000-6

Delmar
5 Maxwell Drive
Clifton Park, NY 12065-2919
USA

Cengage Learning is a leading provider of customized learning solutions with office locations around the globe, including Singapore, the United Kingdom, Australia, Mexico, Brazil, and Japan. Locate your local office at: **international.cengage.com/region**

Cengage Learning products are represented in Canada by Nelson Education, Ltd.

For your lifelong learning solutions, visit **delmar.cengage.com**

Visit our corporate website at **cengage.com**

Notice to the Reader

CONTENTS

PREFACE

OBD II: Functions, Monitors, & Diagnostic Techniques has been designed to fill the void between a text covering the theory of computer-controlled vehicles and one covering advanced engine performance. The design of this text revolves around common Onboard Diagnostics Second Generation (OBD II) problems. The emphasis is on understanding how the OBD II system functions and the use of the onboard monitoring system in repair and maintenance. Since the monitoring system is a frequent source of problems, the text addresses the problem of monitors not set to complete during normal driving and how to get them set either prior to or after a repair. You will find frequent references to states that have emission testing programs and the reasons why specific vehicles are either rejected of failed. As such, this text should be a useful tool for the technician working in an emissions testing area of the country, since there is frequent reference to failed or rejected vehicles.

However, the text is not solely written for emission program technicians. Instead, it assumes that any technician should follow specific techniques when dealing with an OBD II vehicle. OBD II is a computer system built into every current production vehicle, and whether or not the vehicle will be emission tested may have little to do with the reasons why the customer brought it into a repair center. The OBD II system is responsible for the monitoring of many of the systems that move the vehicle down the street. A technician cannot ignore the OBD II system in the repair process. Customers will not accept a vehicle with the check engine light on after repair. The check engine light has been designed to inform the customer that there is a malfunction within the system. Technicians need to understand the importance of this light and just how it is controlled to be able to effectively repair an OBD II vehicle.

When OBD II was introduced, the most frequent problem was a lack of understanding of the monitoring system: how it functioned, why it might not function, and how important it was for the overall function of the system. Although OBD II has been in existence since 1996, there has been little written to help the student or technician understand each section and how it functions. Of particular importance are the monitoring system and its function in the setting of diagnostic trouble codes (DTCs). In my teaching career of 40+ years I have found that technicians do not normally have difficulty repairing a vehicle that actually generated a DTC or multiple DTCs. Even back in the early days of computerized vehicles, technicians read the DTC, repaired the vehicle, and cleared the code. OBD II requires that the technician modify his "clear the code and see if it comes back" mentality. Clearing codes sets all of the monitors into a not-run state that may take a week of normal customer driving to reset. This text will include recommendations to not clear the code all of the time and also not give a customer back his vehicle until all monitors have been set to a run status. The job of the technician, during the repair process, will be to set the monitors to a run status and not have any additional DTCs show up prior to returning the vehicle to the customer. The technician must retain control of the vehicle and set the monitors to a run status.

We will also look at additional techniques to aid in the repair and maintenance of an OBD II vehicle. Mode 5, mode 6, DSOs, and scanner use are covered in detail and will be helpful to any instructor or technician trying to understand how the OBD II system can be diagnosed with accuracy. The text also looks at the impact that controller area networks (CANs) have had and will continue to have on the diagnosis and repair of modern vehicle systems.

This text has been written in an easy, straightforward manner to simplify the understanding of OBD II. It can stand alone in an OBD II class or be used in addition to a computer-controlled textbook. It is the backbone of several OBD II seminars designed for technicians. As such, it will serve as a ready reference for technicians out in the field. We at Cengage hope it will complement other texts that are part of our extensive automotive library. Thanks for using it.

ACKNOWLEDGEMENTS

When I sat down to write this text, I had no idea how much information I would have to put together to make it a complete and comprehensive treatment of OBD-II. I realized early on that I would need the help of individuals and support of family and friends. My wife Carol has been supportive of my writing even when it meant not doing something together because a deadline was approaching. We celebrated 40 years together this past year. They have flown by faster than I ever would have thought possible and I love her more now than ever.

As usual, I have enlisted the help of some specific individuals. Mike Gustafson formally from Vetronix and Bosch has always been helpful, and I appreciate his help and friendship. My friend Jim Wellman from Envirotest acted as a sounding board for some concepts. Ken Zanders, one of the best automotive instructors ever was very helpful in the preparation of the chapter on CAN and I thank him. Ken allowed me to use some of his DSO patterns and a couple of photos. He is an instructor's instructor and one of the best around. I value his friendship. I also appreciate the helpful reviews that were done prior to putting the text into production. Their suggestions were appreciated.

I even enlisted the help of my granddaughter Alexandra Robey (Alex) in the cut and paste end of text preparation. She did a great job! Thanks. There were times during the preparation of this text that I enlisted the help of my son, Keith. He is a high school automotive instructor of the highest caliber. He looked at some of the material and also allowed me shop time when I needed it.

I sincerely hope that this text will help all of you either involved in the teaching of technicians or technicians yourself. Please contact me with suggestions for the next edition. Helpful suggestions are always welcome.

Al Santini

REVIEWERS

The author and publisher would like to thank the following instructors for their invaluable feedback:

George Generke
College of DuPage
Glen Ellyn, IL

Steve Tomory
Rio Hondo College
Whittier, CA

Alex Wong
Sierra College
Rocklin, CA

ABOUT THE AUTHOR

Al Santini has been involved in automotive education for over 40 years, starting with teaching high school, vocational school and community college. For the past 9 years, he has run technician update seminars, primarily for State Emission programs.

Al has been married to his wife, Carol, for 41 years. They have two adult children, Amy and Keith. Amy is married to Jeff Robey and they have two children, Alex, who is 8 years old and Tyler, who is 4 years old. Keith, is an automotive instructor and department chairman at Addison Trail High School. Keith is married to Eileen, and they have one child, Jake, who is 4 years old.

Al has authored other automotive textbooks: Automotive Electricity and Electronics (three editions), Tech One Automotive Electricity and Electronics, the L1 prep guide for Mitchell Manuals, and numerous articles. He is active in the North American Council of Automotive Teachers (NACAT), and the Illinois College Automotive Instructors Association (ICAIA). Al is ASE certified and has been involved in NATEF certification since its inception.

When he is not teaching, Al enjoys sailing, camping and bike riding.

INTRODUCTION TO OBD II

OBJECTIVES

At the conclusion of this chapter you should be able to: ■ List the major differences between OBD I and OBD II ■ Identify the year that OBD II became mandatory ■ Identify the location and function of the DLC ■ Understand the function of the monitoring system ■ List the various FTP standards for OBD I and OBD II ■ Be able to read and understand various emission testing forms

INTRODUCTION

The term OBD (On-Board Diagnostics) refers to systems of self-diagnostic and reporting capability on modern vehicles. OBD I was placed on vehicles in the early 1980s. At that time, manufacturers began installing processors to handle fuel and ignition functions on vehicles. Each year brought more and more functions to the onboard processors. Beginning in 1996, the *EPA (Environmental Protection Agency)* mandated that all vehicles follow the *California Air Resources Board (CARB)* recommendations for emission control. The "new" system would be called OBD II, or On-Board Diagnostics, Second Generation. Prior to 1996, OBD I was primarily an engine management system. After 1996 it became an emission testing strategy that was standardized from manufacturer to manufacturer. The standardization was welcomed by the automotive industry. The maze of different diagnostic connectors was replaced with one 16-pin connector called the *data line communication (DLC),* which is generally in the same place on most vehicles (Figure 1-1). In this chapter we will examine the basics of OBD II and look at its function as an emission testing strategy.

A standardized protocol for communication was established called *generic communication.* In addition, all vehicles would generate *diagnostic trouble codes (DTCs)* that would be the same from vehicle to vehicle. Freeze frame data would be available for at least one DTC, and a check engine light (Figure 1-2) would be illuminated on the dash to indicate that a DTC was present.

This was a radical departure from OBD I, where each manufacturer could do whatever they wanted, generating codes that were different and using a

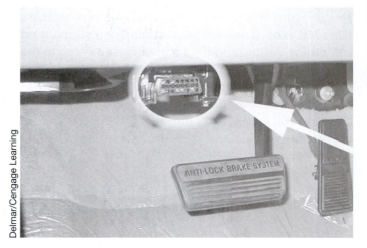

Figure 1-1 The DLC is usually found under the driver's side of the dash.

MIL

Figure 1-2 The check engine light, or MIL, informs the consumer of a problem.

connector placed anywhere on the vehicle that required special adapters to allow communications.

1996 AND OBD II

OBD II originated in 1996 and continues to evolve as an emission and engine management system. True OBD II is an emission testing strategy, offering a simple method of using a scanner to determine problems within the emission system. Many states use this function to pass/fail vehicles produced after 1996.

Emission tests done on vehicles prior to 1996 (Figure 1-3) generally involved a tailpipe sample test using a *dynamometer* and an *infrared exhaust analyzer.* Vehicles were tested and failed for the presence of hydrocarbons (HC), carbon monoxide (CO), and sometimes oxides of nitrogen (NO_x). The method of testing an OBD II vehicle is simpler and easier to accomplish than an OBD I vehicle. The programming within the vehicle processor has the ability to analyze sensor inputs and outputs and, after running a series of tests called *monitors*, can turn on the *malfunction indicator lamp (MIL).* If the MIL is on, it means that a monitor or test has found something wrong, and if a scanner is connected to the DLC, it will indicate a DTC or trouble code in addition to the conditions that were present when the DTC was set. This set of conditions is called *freeze frame.* In this book, we will look at all of the OBD II functions including monitors, DTCs, and freeze frame functions.

WHY OBD II?

OBD I evolved into OBD II because the earlier system of engine management did not allow for any internal emission testing (Figure 1-4).

With OBD I, states had to sample the tailpipe emissions to conduct pass/fail testing. OBD II's proactive shift to emission testing strategy allows the vehicle to test itself and immediately notify the consumer of a problem by turning on the MIL. OBD II does include engine management, but it is primarily an internal

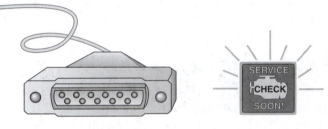

On-Board Diagnostic OBD II

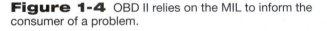

A shift from exhaust emissions to pollution prevention

Figure 1-4 OBD II relies on the MIL to inform the consumer of a problem.

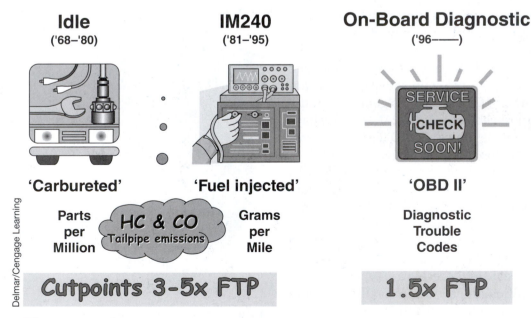

Challenges to Repair Industry

Idle ('68–'80)	IM240 ('81–'95)	On-Board Diagnostic ('96——)
'Carbureted'	'Fuel injected'	'OBD II'
Parts per Million	Grams per Mile	Diagnostic Trouble Codes
HC & CO Tailpipe emissions		
Cutpoints 3-5x FTP		1.5x FTP

Figure 1-3 OBD II vehicles turn on the MIL at 1.5 times the FTP.

Delmar/Cengage Learning

Figure 1-5 Kiosk set up to do an OBD II emission test.

emission testing strategy. It allows a state testing facility to "plug in" and scan the vehicle to determine pass/fail status. OBD II allows for the elimination of tailpipe testing by running the vehicle on *dynamometers* at various loads and speeds. It thus greatly simplifies the emission test. In fact, it allows for some kiosk testing (Figure 1-5), where the vehicle owner plugs the vehicle in and determines pass/fail status with no state employee present. It is a simple straightforward plug-in that scans for monitors, DTCs, and MIL status. It takes less than five minutes, and with every vehicle having the same connector located approximately in the same place, the test is less expensive to administer.

In this text, we will look at the emission test as administered by a typical U.S. state, utilizing the procedures and forms from the State of Illinois. We will examine how the test is administered and follow failed vehicles as they get repaired and retested. We will use various state forms that give a view into the history of the vehicle as well as the specifics of the individual test.

Figure 1-6 shows the history of a 1996 GMC. Notice that on April 3, 2000, the vehicle was given an IM-240 emission test. It passed the test in both 2000 and in 2002.

Vehicle Info and Test History

VIN   Plate   VIR **V006526606** Enf **SOS ENF**   **P**
Obs   XLck **N** HExt **Y** Vol **N**   OBD-ONLY **N**
Year **1996** Make **GMC**   Model **JIMMY** Cyl **6**   Type **1** Adap   Asn **04/2004**
Status **SELECTED FOR SOS ENFORCEMENT ON 08/07/2004**   Veh# **3551704**   Exp **07/2004**
Own   DLN   Reg **11/2004**
Own   DLN
Addrs   Zip   Cty **82**
Flt **N**   OC **2**   Des **008**   Pur **11/15/2000**   Upd **02/05/2004**   Ta **9**   Clr **No Date**   **N**
Pending suspension on: **10/20/2004**

Action	Vir#	Stn	Ln	Date	Time	Test Type	Test No.	Test Mode	Test Rslt	E C O	Rsn
Test	V010458147	34	1	08-12-2004	08:43	OBD	1	O	Reject	- P R	OM
Test	V010370123	32	1	08-05-2004	11:38	OBD	1	O	Reject	- P R	OM
Test	V010463762	34	2	08-05-2004	09:53	OBD	1	O	Reject	- P R	OM
Test	V010470647	34	3	06-22-2004	15:05	OBD	1	O	Reject	- P R	OM
Test	V010455659	34	1	05-06-2004	13:51	OBD	1	O	Reject	- P R	OM
Test	V010464958	34	2	04-30-2004	14:55	OBD	1	O	Reject	- P R	OM
Test	V006526606	34	2	04-26-2002	09:09	240	1	O	Pass	P P P	
Test	V001633968	34	1	04-03-2000	13:51	240	1	O	Pass	P P P	
Init	Z086515758	88	0	02-27-2000	12:38					.	I
Unexempt	P113359565	80	0	12-20-1998	11:56						S
eXempt	P108405030	80	0	07-04-1996	17:31	T					Z

Delmar/Cengage Learning

Figure 1-6 Vehicle history covering eight years.

When the next biennial test came up, the vehicle was slated for an OBD test. It was rejected for the next six tests, and the owner eventually lost his driver's license. The next state form that we will use within this text, the *Repair Diagnostic Report (RDR)* (Figure 1-7), is for the same GMC, but includes the reason for the reject.

The bottom section lists a series of "not ready" monitors under the status column. This is the readiness status that is so much a part of OBD II. Readiness is a way of saying whether the vehicle has tested itself under normal driving conditions. Each of the systems listed under the description column requires a set of tests to be done to evaluate a part of the OBD II system. For instance, the first one listed is for catalyst efficiency status. The series of tests that will determine the catalyst's efficiency is the CAT (catalytic converter) monitor, and in the case of this GMC, it has not been run. Because the majority of the monitors are listed as not ready, it means that the OBD II system has not been tested. The only way to determine overall status, generate DTCs, and save freeze frame data is to run the monitors. Rejects caused by insufficient monitors run is one of the greatest challenges to the modern technician. If the monitors will not run, there is no way that the MIL or check engine light can inform the driver of a problem involving the emission system. The check engine MIL is only functional when the monitors run.

The entire system is based on the running of the monitors. Only after running the monitors does the system have the ability to generate DTCs and freeze frame data and turn on the MIL as Figure 1-8 shows.

Throughout this text, we will examine the function of the monitors in great detail. Without the monitors running to completion, the system has no way of determining if all is well. In many states that are doing mandatory emission testing, lack of monitors results in more rejects than failures.

OBD - REPAIR DIAGNOSTIC REPORT (RDR)

[View Test Details]

This Report contains information that will help a repair technician diagnose and repair your vehicle.

Plate: VIN: VIR: V010458147 Make: GMC Model: JIMMY Model Year: 1996	Test Time: 08/12/2004 08:43:59 Test No: 1 Vehicle Type: LDT1, U Test Type: OBDII Station: 34 Lane: 1	MIL Commanded: OFF Bulb Check: PASS Ready Status: NOT READY

OBD TEST RESULT - REJECT

We were unable to complete testing your vehicle today because the check of OBD readiness monitors indicates that the vehicle has not completed all required emissions component evaluations. The results of the OBD systems readiness check are listed below. Systems/components preventing the completion of the OBD test are listed as NOT READY. This can happen if you have recently had repairs performed on your vehicle, replaced the battery, or if the battery has run down or been disconnected.

Your vehicle did not fail the OBD test and this reject does not mean that there is anything wrong with your vehicle.

In order to prepare most vehicles for the OBD test, the vehicle should be driven for several days under a variety of normal operating conditions. This includes a mix of highway and stop and go, city type driving.

Your vehicle owner's manual may provide more specific instructions on preparing the vehicle for the OBD test. Additional information on preparing the vehicle for testing can be provided by consulting your dealer service department or independent repair facility.

CODES	DESCRIPTION	STATUS
CAT	Catalyst Efficiency Status:	NOT READY
CATHEAT	Catalyst Heating System Status:	Not Supported
EVAP	Evaporative System Status:	NOT READY
AIR	Secondary Air System Status:	Not Supported
AC	Air Conditioning Refrigerant Status:	Not Supported
O2S	Oxygen Sensor System Status:	NOT READY
OS2HEAT	Heated Oxygen Sensor System Status:	NOT READY
EGR	Exhaust Gas Recirculation System Status:	NOT READY

Figure 1-7 Repair Diagnostic Report for a single emission reject.

OBD II in a nutshell

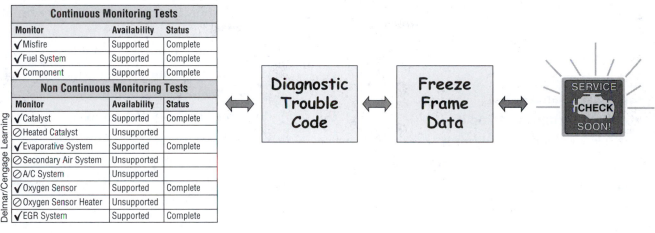

Continuous Monitoring Tests		
Monitor	**Availability**	**Status**
✓ Misfire	Supported	Complete
✓ Fuel System	Supported	Complete
✓ Component	Supported	Complete
Non Continuous Monitoring Tests		
Monitor	**Availability**	**Status**
✓ Catalyst	Supported	Complete
⊘ Heated Catalyst	Unsupported	
✓ Evaporative System	Supported	Complete
⊘ Secondary Air System	Unsupported	
⊘ A/C System	Unsupported	
✓ Oxygen Sensor	Supported	Complete
⊘ Oxygen Sensor Heater	Unsupported	
✓ EGR System	Supported	Complete

Diagnostic Trouble Code ⟷ Freeze Frame Data ⟷ SERVICE CHECK SOON!

Delmar/Cengage Learning

Figure 1-8 OBD II functions only after the monitors have run.

CHALLENGES TO THE REPAIR INDUSTRY

Many states began testing vehicle emissions over 20 years ago. Initial testing of 1960s and '70s era vehicles was generally done by measuring HC and CO at idle and comparing the levels actually produced against the standards of the *Federal Test Procedure (FTP)*. Typically a vehicle failed if it exceeded the FTP standards by about five times. Another way of saying this is that a vehicle would fail the emission test if tailpipe emission were five times, or 500% of, the standard. When OBD I arrived on the scene in the 1980s, the emission test reflected the fact that the vehicles were now typically computer or processor controlled, with the majority of vehicles being fuel injected. These vehicles are generally tested on dynamometers and are failed if the actual emissions coming out the tailpipe exceed three times the FTP. You can see that the testing standards are becoming stricter, moving down from five times the FTP to three times the FTP. OBD II was introduced as a radical departure from typical vehicle emission control. No longer would there be a need to actually sample the exhaust out the tailpipe or run the vehicle on a dynamometer. Instead, the vehicle would test itself (run monitors) and turn on the MIL if conditions were present that could result in emissions 1.5 times greater than the FTP standard. It's easy to see that OBD II is a tighter test that any other emissions test. The standards for pass/fail have been reduced from five times to 1.5 times the FTP, which is a substantial reduction. In addition, the actual standards of the FTP have also been reduced. The challenge to the repair industry comes in recognizing that the MIL is illuminated at 1.5 times the FTP and that they will probably not be able to measure this small of an increase out the tailpipe with typical garage infrared analyzers. The emphasis now must be on utilizing the scanner as well as the usual diagnostic fuel injection equipment, like a *digital storage oscilloscope (DSO)*. Additionally, technicians must recognize the issue of running the monitors prior to getting a fixed vehicle emission tested. Most states will either reject or fail a vehicle that has not run sufficient monitors. We will spend extensive time on monitors in this text, so if you are a bit confused as to their function, do not worry. It will be clear by the time we are done.

CONCLUSION

In this chapter we introduced OBD II. You should now understand that it is an emission testing strategy that relies on the monitors being run to completion. The monitors will generate DTCs and save freeze frame data for the technician to look at and analyze. Various states use the OBD II system as their method of emission testing vehicles produced from 1996 onward. Once the monitors have run, the state can look at the results and either pass or fail the vehicle. If the check engine light is on, most states will fail the vehicle. However, if the monitors will not run, most states will reject the vehicle.

REVIEW QUESTIONS

1. Technician A states that OBD II really began with 1996 vehicles. Technician B states that OBD II was patterned after California Air Resources Board requirements. Who is correct?
   a. Technician A only
   b. Technician B only

c. Both Technician A and B

d. Neither Technician A nor B

2. Technician A states that OBD I was primarily an engine management system. Technician B states that OBD II is primarily an emission testing strategy. Who is correct?

a. Technician A only

b. Technician B only

c. Both Technician A and B

d. Neither Technician A nor B

3. OBD II standardized

a. location of the DLC

b. connections within the DLC

c. communications

d. all of the above

4. Technician A states that the MIL will be illuminated when the PCM (powertrain control module) runs a monitor. Technician B states that monitors must run before any DTCs will be detected. Who is correct?

a. Technician A only

b. Technician B only

c. Both Technician A and B

d. Neither Technician A nor B

5. A reject will occur at a state emission test center if a vehicle

a. has DTC's present

b. has the check engine light on

c. is beyond the time limit for the test

d. has insufficient monitors

6. Technician A states that a standardized communications protocol will allow any OBD II scanner to retrieve manufacturer-specific DTCs. Technician B states that the generic protocol will generate limited scanner information that can be used to pass or fail an emission test. Who is correct?

a. Technician A only

b. Technician B only

c. Both Technician A and B

d. Neither Technician A nor B

7. "Not ready" is indicated during an emission test. Technician A states that this indicates that the check engine light is on. Technician B states that this indicates that DTCs are present. Who is correct?

a. Technician A only

b. Technician B only

c. Both Technician A and B

d. Neither Technician A nor B

8. Freeze frame data is present when

a. a monitor runs

b. the vehicle is emission tested and passes

c. a DTC is stored

d. a technician clears the memory of the PCM

9. A check engine light is on. This indicates that

a. a monitor has set a DTC

b. a monitor has run and not found anything wrong

c. the grams per mile of pollutants out the tailpipe has exceeded five times the federal limits

d. a technician has cleared all DTCs

10. The MIL is off and a DTC remains when the

a. monitor runs again and does not detect the same problem

b. monitor has not run to completion

c. freeze frame has been cleared

d. DTC has been cleared with a scanner

OBD I VERSUS OBD II

OBJECTIVES

At the conclusion of this chapter you should be able to: ■ Identify the input sensors common to OBD I ■ Identify the output sensors common to OBD I ■ Identify the input sensors common to OBD II ■ Identify the output sensors common to OBD II ■ Decode an OBD II DTC ■ Identify the enabling criteria for an OBD II monitor

INTRODUCTION

In this chapter, we will compare the OBD I system to OBD II. The clearest differential will be the year, since the system is referred to as OBD I prior to 1996 and OBD II in the years afterward. Additionally, we will look at the engine management side of OBD I versus the emission testing side of OBD II. We will look at the function of the oxygen sensor as it relates to both OBD I and OBD II plus additional inputs and outputs.

OBD I

It's important to realize that most functions of the OBD I system are accomplished or duplicated within the OBD II system. The vehicle has to have engine management or it will not function. So let's look at what OBD I does. In Figure 2–1, engine management is easily broken down into fuel, ignition, and frequently, engine cooling.

The same sets of sensors are generally utilized for these three functions. Remember that OBD II will typically utilize the same sensors. Getting the vehicle started is normally accomplished through the use of an engine coolant temperature sensor, an intake air temperature sensor (ambient or outside temperature), some type of a crank sensor that can measure engine speed, plus some method to look at the amount of air actually going through the engine. Measure the temperature, the engine speed, and the amount of air, and then the vehicle processor has enough information to get the engine started. Look at the block diagram in Figure 2–1, which shows the input sensors on the left, the processor in the middle,

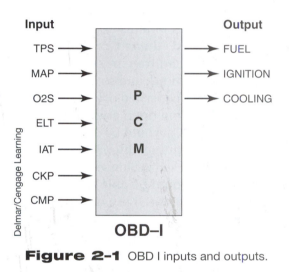

Figure 2-1 OBD I inputs and outputs.

and the outputs on the right. The most important output here is fuel injection because the correct amount of fuel has to be delivered or the engine will not start. Most OBD I vehicles had distributors, so initial timing was not a concern. The position of the distributor determined when the spark was delivered. This is called *base* or *initial timing*.

Once the engine starts, additional sensors come into play and additional processor outputs begin. The timing is altered and controlled by an electronic spark advance circuit. As engine speed and load change, a signal to the electronic ignition module advances or retards the timing. Once the engine is up to operating temperature, the processor will turn the cooling fan on and off as needed to control the maximum coolant temperature. The block diagram shows these additional outputs on the right.

Probably the most important sensor, once the vehicle is running, is the oxygen sensor (O2S). The job or function of the O2S is to measure the oxygen content of the exhaust and send this information as an input to the processor. During the OBD I era, the phrase *open loop* or *closed loop* came into use. Figure 2–2 shows the closed loop where the signal from the O2S is routed to the processor, and the processor uses the signal to vary the amount of fuel injected.

As the fuel burns, it generates another O2S signal. This loop, which involves burning fuel, measuring oxygen in the exhaust, and then adjusting injection, occurs over and over many times a minute. Closed loop means that the processor is using the signal from the O2S. Open loop means that the signal is not being used. Open loop occurs during start-up, early warm-up, and on some vehicles during heavy acceleration. At all other times closed loop is functioning and is utilized to vary the fuel injection pulse for maximum economy, adequate power, and minimum emissions. Figure 2–1 contains an abbreviated list of the inputs and outputs that a typical OBD I vehicle had. Some vehicles had transmission control, EGR (exhaust gas recirculation), and so on.

OBD I diagnostics consisted of a few trouble codes oriented toward finding open and/or short circuits. The typical list of DTCs would be 20 to 50 items long with virtually no output testing. For example, if the TPS (throttle position sensor) developed an open circuit, the processor would be able to "see" the open, turn on the check engine light, and possibly go to some backup mode that would allow the vehicle to continue moving. The driver was informed by the light that something was wrong. As long as the circuit was open, the check engine light would remain illuminated. If the short disappeared, the light would go out. Repairing these vehicles was usually done by "clearing the code to see if it would come back." In this manner a technician could identify an intermittent problem by observing the light, clearing the code, driving the vehicle, and again observing the light. With OBD I, it was common practice to clear the code to see if it came back. It is important to note that this procedure is not acceptable in all OBD II DTCs. We will examine the correct procedure when we cover diagnostics.

Let's go back to our hypothetical OBD I and look at what the system does if an output is open. Imagine an open fuel injector creating a dead cylinder. Here is where OBD I fell flat—it did not recognize the open fuel injector, nor did it see the dead cylinder. In most cases outputs were ignored from a diagnostic standpoint. Inputs were tested for opens and shorts only. OBD I also did not have the ability to look at a sensor in terms of its accuracy. A coolant temperature sensor that read 160 degrees when the coolant temperature was actually 200 degrees would not be found. Technicians would have to wait until OBD II to get output testing and diagnostics plus rationality testing.

WHAT DOES OBD II DO?

OBD II's first major addition is the ability to test outputs (Figure 2–3). That, plus the potential for thousands of DTCs, makes the system very flexible.

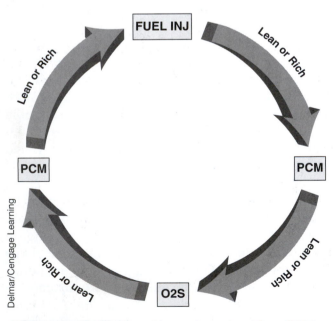

Figure 2-2 Closed loop where signal from O2S is used to adjust injection.

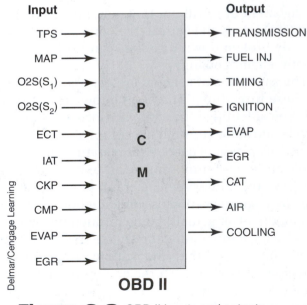

Figure 2-3 OBD II inputs and outputs.

Also, remember that the definition of a DTC will be the same for all manufacturers. For example, a P0201 signifies that the PCM, or powertrain control module (the "P" in the DTC), has detected a fuel injector circuit problem with cylinder #1 during the routine generic testing (the "0" in the DTC). The full description follows in Figure 2–4.

The other consideration is that a P0201 on a Jeep means the same thing that it does on a Mustang or a Firebird. The generic notation (the 0 after the P) indicates that the SAE (Society of Automotive Engineers) has defined the DTC and standardized its description. OBD II allows the manufacturers to still have their own diagnostic trouble codes (Figure 2–5) and diagnostic definitions; however, that would be noted by a P1 rather than a P0.

Another function that we see in OBD II is that of system testing during normal driving. This system of monitoring the various emission systems is utilized within most State emission programs. It is important to note that the DTC, the MIL and freeze frame are all functions of the monitoring system. If a monitor will not run, for whatever reason, then a DTC cannot be generated, the MIL cannot be turned on, and without a DTC there can be no freeze frame. This system of monitors is divided into two sections: the continuous monitors and the non-continuous monitors. Remember that we said that OBD I functions were carried over into OBD II. Within the continuous monitor section is the comprehensive component monitor. This monitor will test the components, both inputs and outputs, and will function most like the diagnostic system of OBD I. If, for instance, the TPS fails in either OBD I or II, it will generate a DTC, although OBD II will have more DTCs for a greater degree of testing. This is the carryover system between OBD I and OBD II. We will look in greater detail at the various monitors in future chapters.

ENABLING CRITERIA

This is a good place to introduce the enabling criteria. Simply stated, the enabling criteria are the set of conditions that need to be present to have the PCM run the monitor. This is one of the distinguishing characteristics of OBD II. The criteria might be very simple like engine running, or very complicated involving driving under various loads, speeds, and temperature. They might involve key on or off time plus fuel levels. It is extremely important to understand that the monitors will not run unless the enabling criteria have been met. They are the heart of the OBD II system, so, as a technician, you must have access to the enabling criteria or you may not be able to get the monitors to run. Let's look at the criteria for a simple monitor (Figure 2–6): an EGR monitor for a 2004 Ford. They consist of six steps preceded by "conditions."

P0201	Cylinder 1 Injector Circuit/Open

Figure 2–4 P0201 Cylinder #1 injector circuit/open.
Delmar/Cengage Learning

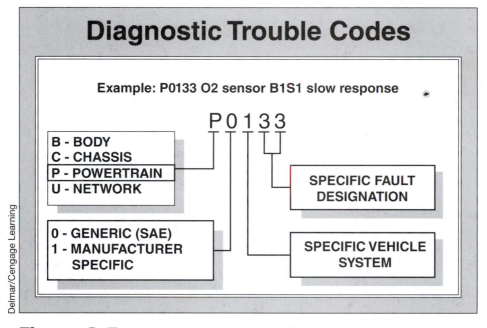

Figure 2–5 Diagnostic trouble codes can be decoded.

Conditions	1. Fuel tank should be ½ to ¾ full; ¾ full is preferred. 2. Operating the throttle smoothly during cruise will minimize the time required for monitor completion.
Step 1	Connect a scan tool to the data link connector. Turn the key ON with the engine OFF. Cycle the key OFF then ON. Clear all DTC's.
Step 2	Monitor the following PID's ECT, EVAPDC, TP. Start the engine without returning the key to off.
Step 3	Idle the engine for 15 sec, then drive at 40 mph until the ECT is at least 170 degrees F.
Step 4	From a stop, accelerate to 45 mph at ½–¾ throttle.
Step 5	Repeat step 4 three times.
Step 6	Check monitor status.

Figure 2-6 Enabling criteria for an EGR monitor.

This is probably one of the simpler monitor drive cycles, but it will take some analyzing to understand. Let's take a step at a time and try to see what is happening to the enabling criteria. Remember the enabling criteria are the set of conditions required for the monitor to run.

1. **Fuel tank should be half to three-quarters full; three-quarters full is preferred.** What does this mean? On the surface it is telling you how much gas needs to be in the tank if the EGR monitor is to run. Think about your customer who tends to run the vehicle with less than half a tank of gas. Will the monitor run? Probably not. This also means that it is likely that an EGR failure would go undetected because the monitor will not run. As we said earlier, monitors need to run to set DTCs. We also need to read between the lines because this also means that the vehicle's fuel gauge must be functional.

2. **Operating the throttle smoothly during cruise will minimize the time required for monitor completion.** This is self-explanatory if you think about it. If we are on the gas and off the gas, the fuel will slosh around in the tank and make the monitor take longer. So using the cruise control is a good idea.

Now, on to the actual enabling criteria drive cycle.

3. **Connect a scan tool to the data link connector. Turn the key ON with the engine OFF. Cycle the key OFF and then ON. Clear all DTCs.** There are a few important things to note here. First: connect a scan tool. The reason for this is that we need to clear any DTCs and know when the monitor runs. Second: cycle the switch in a specific manner. Third: clear any DTCs. This is a good time to discuss the clearing of any DTCs. Why would the manufacturer require this? Simply, if the PCM has figured out that something is wrong and set a DTC, it will likely prevent additional monitors from running. Most non-continuous monitors will require clearing DTCs prior to running. Remember this when the customer comes into the shop with the MIL (check engine light) on. The odds are good that other monitors have not been running since the light was activated. This is important because when we fix the DTC and clear the code, the other monitors will probably run and may detect additional problems. At that point, the customer should be told that once the current DTC is fixed, other DTCs may appear as the rest of the monitors are run. Customers need to know that the current DTC might not be the only problem. Getting the monitors to run should become an integral part of every OBD II repair.

4. **Monitor the following PIDs: ECT (engine coolant temperature), EVAPDC, TP. Start the engine without returning the key off.** This step is very important because it tells the technician that after clearing the DTC, he or she should start the engine without turning the ignition key off. This sequence starts the monitor.

5. **Idle the engine for 15 seconds; then drive at 40 mph until the ECT is at least 170 degrees F.** Let's think about this seemingly simple step. Idle and then drive until the engine warms up to at least 170 degrees F. By reading between the lines we see that we need to have a functional and accurate ECT that can read 170 degrees of engine heat. Place the vehicle in one of the northern states during winter and the importance of a functioning thermostat is highlighted. Don't forget to read between the lines.

6. **From a stop, accelerate to 45 mph at one-half to three-quarters throttle.** Again, reading between the lines means that this vehicle needs to have a functioning speedometer. We also must perform an aggressive acceleration from a stop.

7. **Repeat step 4 three times.** Stop and accelerate three more times for a total of four from a stop to 45 mph. The obvious question is where do you do this? If your shop is in a city, it might be physically impossible without getting a ticket or at least annoying other drivers.

The above steps should allow the monitor to run to completion. Remember, running the monitor is the prerequisite to figuring out what is wrong or right with the vehicle.

We will spend more time looking at enabling criteria in future chapters.

CONCLUSION

In this chapter, we have spent substantial time looking at the differences between OBD I and OBD II. We saw how the inputs have generally remained the same while the outputs have evolved. We examined how the DTCs are generated within OBD I and how they are generally handled by a technician. We also discussed decoding a DTC and featured examples of the enabling criteria necessary for monitors to run.

REVIEW QUESTIONS

1. Technician A states that OBD I was an emission testing strategy beginning in 1996. Technician B states that OBD II was an engine management strategy used prior to 1996. Who is correct?
   a. Technician A only
   b. Technician B only
   c. Both Technician A and B
   d. Neither Technician A nor B

2. OBD I systems
   a. controlled all engine outputs directly
   b. utilized a large number of inputs but had limited outputs
   c. did not have distributors
   d. used closed loop to control engine temperature

3. Technician A states that the oxygen sensor signal is used to determine adjustments to fuel. Technician B states that closed loop operation refers to the status of the oxygen sensor signal. Who is correct?
   a. Technician A only
   b. Technician B only
   c. Both Technician A and B
   d. Neither Technician A nor B

4. Clearing the code to see if it comes back is a procedure useful on
   a. OBD I
   b. OBD II
   c. closed loop
   d. monitor setting

5. Technician A states that OBD II generic codes are generated when monitors run. Technician B states that DTCs can be "decoded" to reveal the system at fault. Who is correct?
   a. Technician A only
   b. Technician B only
   c. Both Technician A and B
   d. Neither Technician A nor B

6. Enabling criteria are being discussed. Technician A states that it is the driving conditions that must be present to get the monitor to run. Technician B states that if a monitor will not run, it is possible that the enabling criteria have not been met. Who is correct?
   a. Technician A only
   b. Technician B only
   c. Both Technician A and B
   d. Neither Technician A nor B

7. The sequence of driving conditions to generate DTCs usually is
   a. very specific
   b. involves driving
   c. called the enabling criteria
   d. all of the above

8. A vehicle's enabling criteria call for an eight-hour soak. Technician A states that this means the ECT and IAT (intake air temperature) must be equal. Technician B states that the PCM must see a time period of at least eight hours to satisfy the soak requirement. Who is correct?
   a. Technician A only
   b. Technician B only
   c. Both Technician A and B
   d. Neither Technician A nor B

9. Technician A states that the enabling criteria should be accomplished during normal driving. Technician B states that individual pieces of the enabling criteria must be followed in sequence or the monitor may not run. Who is correct?
   a. Technician A only
   b. Technician B only
   c. Both Technician A and B
   d. Neither Technician A nor B

10. A monitor will not run. Technician A states that this is normal on a vehicle that has no real problems. Technician B states the repair is not complete until the monitors run. Who is correct?
    a. Technician A only
    b. Technician B only
    c. Both Technician A and B
    d. Neither Technician A nor B

DIAGNOSTIC TROUBLE CODES (DTCs)

OBJECTIVES

At the conclusion of this chapter you should be able to: ■ Determine what a DTC is ■ Recognize the importance of running monitors ■ Decode a DTC ■ Analyze a failed emission report ■ Analyze a reject emission report ■ Analyze code setting criteria ■ Scan and understand the function of the MIL ■ Scan a vehicle for readiness status

INTRODUCTION

In this chapter, we will examine the specifics of OBD II DTCs, how they are generated, what they mean, and how they are cleared. Additionally, we will look at some examples of State emission test forms. Code setting criteria will be discussed with emphasis on using the scanner to look at readiness status and MIL function.

Before we get into the actual generation and repair of DTCs, it is important to again refer to the monitoring system and enabling criteria that we touched on in the previous chapters. A distinguishing feature of the OBD II system is its ability to generate DTCs. An OBD I system might have a hundred trouble codes, limited mostly to input diagnostics, but it had virtually no output diagnostics available. OBD II, on the other hand, can generate thousands of DTCs and look at both inputs and outputs. Additionally, the system can do comparisons between different inputs to determine if a component is working correctly. For example, by looking at the O2S signal after the catalytic converter and comparing it to the signal of the O2S before the converter, OBD II can generate a converter efficiency DTC. A functioning CAT will generate very specific O2S waveforms as it ages. Eventually, when the converter's efficiency falls below a pre-determined level, the system might generate what is seen in Figure 3-1.

The system also can do rationality checking of inputs. This is a check of an input against another input to see if the signal makes sense. A P0120 is a good example of a rationality check. It basically says that

| P0420 | Catalyst System Efficiency Below Threshold (Bank 1) |

Figure 3-1 A P0420 results when the CAT monitor detects that efficiency has dropped below the standards.
Delmar/Cengage Learning

the TP (throttle position sensor) signal does not agree with the MAP (manifold absolute pressure) sensor. It operates on the assumption that there is an expected change in both signals when the engine is accelerated and that they will be in agreement. Notice that it does not say that the TP is bad or that the MAP is bad. It says that they do not agree. As a technician your job will be to look at both signals and determine which is not correct.

Additionally, there are functionality DTCs. They answer the basic question of how good is the component. A P0155 indicates that the ECT circuit is not functioning correctly, and in all likelihood the signal is being ignored through a process of substitute signal values. We will look at substitute values in a later chapter.

All DTCs fall into three categories: rationality, functionality, and system testing. All DTCs require a monitor to be run. Remember: No monitor means no DTC. There are times when a technician says there is nothing wrong with the vehicle because there are no DTCs. That statement might be correct if all of the monitors have run. We will analyze the monitor issue in detail in the next chapter. For now, just keep in mind that the setting of any DTC is based on a specific monitor running and detecting the problem.

HOW TO DECODE THE DTC

OBD II separates DTC into categories. The first systems had two possibilities: manufacturer's code or a generic code at the powertrain level. The first letter of the code indicates the powertrain if it is a "P" or the body system if it is a "B." We will look at the P codes primarily here. The second number, either a "0" or a "1," indicates which side of the powertrain has generated the code. When OBD II was originally designed, all manufacturers were required to have a functioning generic system. Generic is another way of saying basic and similar. The generic side of the system is the side that is the same on all vehicles. Standardization among the various manufacturers was virtually nonexistent until OBD II. After 1996, all manufacturers were required to have their systems function according to SAE (Society of Automotive Engineers) standards. Within the generic system are the P0 DTCs. All DTCs have the same definitions regardless of the vehicle manufacturer. In addition, any OBD II scanner can be connected using the same connector and will communicate with the vehicle. This in itself is a tremendous improvement over OBD I, where every vehicle manufacturer used different communication protocols and connectors. It is important to note that OBD II did not eliminate the ability of the manufacturer to utilize additional communication; it just required that they have the generic system built into the vehicle. This would allow anyone (primarily a State) to connect and use the system to determine the status of the vehicle. Remember various States

utilize the OBD II generic system to determine pass/fail status of a vehicle. This is well within the capabilities of the system as it was primarily designed as an emission testing strategy.

Let's go back to the DTC. So far we have looked at the first letter and determined that it would specify the origination of the DTC, with P for powertrain and B for body being the most common. The second digit, either a 0 or a 1, would indicate the system within OBD II, with the 0 being generic and the 1 being manufacturer specific. Next come three numbers. The first is used to indicate the vehicle system, and the last two for the specific fault. Overall it looks like Figure 3–2.

The P0133 illustrated can be broken down as a generic powertrain code within the bank 1 sensor 1 oxygen sensor, which appears to have a slow response time. Notice how specific this code is. Not only is the location noted, but the problem with the sensor is highlighted (slow response). Note: A bank code refers to the set of cylinders on a V-type engine. Bank 1 is the set of cylinders that contain cylinder number 1, and bank 2 is the opposite set of cylinders.

Remember that the 0 indicates that the generic part of the system is where the fault resides or originated. If a 1 shows up as the second digit, then the manufacturer's system is at fault. In many cases, the technician needs to look at both sides to analyze the vehicle. When faced with an emission failure, the technician will frequently start on the generic side to duplicate the emission test conditions, move to the

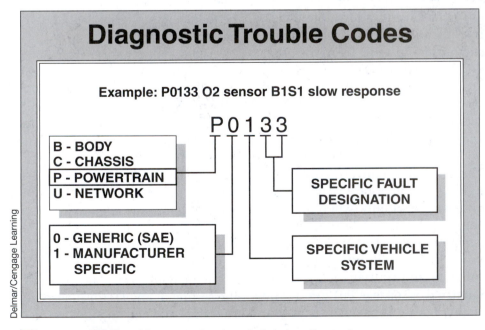

Figure 3-2 DTCs should be decoded during diagnosis.

manufacturer's side to additionally diagnose and repair the vehicle, and finally go back to the generic side to run the monitors. The last step, that of running the monitors, is extremely important for the total repair. Getting the vehicle repaired and retested becomes the total job. Most States scan for monitor status and will reject the vehicle if the monitors have not run. Technicians need to understand that their job is not complete until they get the monitors to run. The monitors are the final test of the repair prior to giving the vehicle back to the customer. As we get deeper into the OBD II system, you will see how this sequence of diagnose, repair, and retest will involve the generic and the manufacturer's side of the

total system. Figure 3-3 shows a fail that generated diagnostic codes. The starting point for the technician will be to duplicate the conditions present during the setting of the code.

Don't forget that vehicles with insufficient monitors will be rejected like the Ford in Figure 3-4. This vehicle did not fail the emission test; it was rejected. With no monitors set to ready, the vehicle probably has something wrong or recently had the battery disconnected or a DTC cleared. The job of the technician will be to first get the monitors to run and then fix any DTCs that pop up. The final test will be to get the monitors to run without generating any DTCs, get the vehicle emission tested, and returned to the customer.

Illinois Vehicle Inspection Report and Certificate

V009992880

The vehicle identified below has been inspected to determine if it is emitting excessive amounts of exhaust and/or evaporative emissions. The inspection consists of an exhaust test for Hydrocarbons (HC) and Carbon Monoxide (CO), or a check of the vehicle's on-board diagnostic (OBD) computer system (1996 and newer OBD-equipped vehicles only) for indication of malfunctions causing or leading to excessive emissions. Also included is a check of the vehicle's gas cap for leaks. CO is a colorless, odorless, toxic gas. HC is unburnt fuel that, when emitted into the air, contributes to the formation of ground-level ozone (smog). Ground-level ozone is a respiratory irritant that causes eye and throat discomfort and damages breathing passages.

If your result is **PASS**, keep the certificate below in your vehicle.

Test Result
FAIL

If your result is **FAIL**, please read the back of this form, the Air Team failure brochure, and see the customer service representative in the station office for more information.

Vehicles produce excessive levels of emissions as a result of poor maintenance, equipment malfunction, or tampering with emissions controls. **We thank you for doing your part to help clean the air we breathe!** If you need further information or assistance, please see the customer service representative in the station office or call: **847-758-3400**

Exhaust Test	REPRINT	N/A		
OBD	HC Hydrocarbons	CO Carbon Monoxide	CO₂ Carbon Dioxide	NOx (Advisory Only) Oxides of Nitrogen
Standards	N/A	N/A	N/A	N/A
Readings	N/A	N/A	N/A	N/A
Result	N/A	N/A	N/A	N/A

Evaporative Test Result	OBD Result On-Board Diagnostic System Check	Retest Code
Gas Cap Test **PASS**	**FAIL - TROUBLE CODES**	**N**

Vehicle Information

VIN:	Model:	S70 GL	Cylinders:	5
Plate :	Year:	1998	Engine Displ:	2 4
Make: VOLV	Odometer:	26000	Trans:	AUTOMATIC
	Class:	LDV, S	Weight:	3000

Test Information

Date: 01/06/2004	Operator IDs:	318753 317609	Test #:	1
Time: 12:49:33	Wait Time:	00:03:07	TFC:	N
Station: 27	Test Type:	OBD	LID:	980397 53.1
Lane: 1	Dyno Settings:		Assigned Date:	01/2004
	Gas Cap Test Code:	A		

Comments
YOUR VEHICLE IS ELIGIBLE FOR TWO RETESTS.

SEE REVERSE SIDE AND CUSTOMER SERVICE REPRESENTATIVE FOR INSTRUCTIONS.

AFTER 12/31/2003, SECOND-CHANCE EXHAUST TESTING IS NO LONGER AVAILABLE FOR VEHICLES FAILING THE OBD TEST.

This test is authorized by the Vehicle Emissions Inspection Law of 1995 (625 ILCS 5/13B). IL 532 2476 VIM 125 DEC. 2001.

Detach on perforation. Keep this certificate in the vehicle at all times.

State of Illinois

VOID

VOID

VOID

Vehicle Emissions Compliance Certificate

Delmar/Cengage Learning

Figure 3-3 A clear fail is one in which the MIL is on and trouble codes are present.

OBD - REPAIR DIAGNOSTIC REPORT (RDR)

This Report contains information that will help a repair technician diagnose and repair your vehicle.

Plate: VIN: VIR: V016401806 Make: FORD Model: WINDST Model Year: 1998	Test Time: 12/12/2005 15:17:17 Test No: 1 Vehicle Type: LDT1, V Test Type: OBDII Station: 21 Lane: 3	MIL Commanded: OFF Bulb Check: PASS Ready Status: NOT READY

OBD TEST RESULT - REJECT

We were unable to complete testing your vehicle today because the check of OBD readiness monitors indicates that the vehicle has not completed all required emissions component evaluations. The results of the OBD systems readiness check are listed below. Systems/components preventing the completion of the OBD test are listed as NOT READY. This can happen if you have recently had repairs performed on your vehicle, replaced the battery, or if the battery has run down or been disconnected.

Your vehicle did not fail the OBD test and this reject does not mean that there is anything wrong with your vehicle.

In order to prepare most vehicles for the OBD test, the vehicle should be driven for several days under a variety of normal operating conditions. This includes a mix of highway and stop and go, city type driving.

Your vehicle owner's manual may provide more specific instructions on preparing the vehicle for the OBD test. Additional information on preparing the vehicle for testing can be provided by consulting your dealer service department or independent repair facility.

CODES	DESCRIPTION	STATUS
CAT	Catalyst Efficiency Status:	NOT READY
CATHEAT	Catalyst Heating System Status:	Not Supported
EVAP	Evaporative System Status:	NOT READY
AIR	Secondary Air System Status:	Not Supported
AC	Air Conditioning Refrigerant Status:	Not Supported
O2S	Oxygen Sensor System Status:	NOT READY
OS2HEAT	Heated Oxygen Sensor System Status:	NOT READY
EGR	Exhaust Gas Recirculation System Status:	NOT READY

Delmar/Cengage Learning

Figure 3-4 A vehicle that has insufficient monitors run will result in a reject.

CODE SETTING CRITERIA

When a DTC sets, it indicates that the diagnostic system has found a problem. When we looked at monitors we used the term *enabling criteria*. The enabling criteria are the various conditions that have to be present for the monitor to run. The code setting criteria are very similar to enabling criteria, except they are for the DTC rather than the monitor. They are a set of conditions that must be occurring and will be used to determine if setting a code is appropriate. Figure 3-5 shows the conditions for a P0121 DTC.

Notice that first it needs to see that there are no "active" TP or MAP DTCs. This is another way of saying the comprehensive component monitor must first determine that the basic function of the TP (throttle position) and the MAP (manifold absolute pressure) sensors are OK. The remaining four conditions indicate what the actual sensor PIDs (parameter identifications) are. PIDs are the individual items that show on a scanner. A PID is a single line of data, and in the case of the P0121, there are four. If these four PIDs match the code setting criteria as listed in the illustration, the MIL (malfunction indicator lamp) or check engine light will illuminate, indicating that a DTC is present.

What Conditions are Present to Set P0121?

- No active TP or MAP DTC
- Engine is running
- MAP sensor reading is less than 55kPa
- Predicted throttle angle is not close to actual throttle angle
- TP sensor angle is greater than 60 percent at 1600 RPM

Figure 3-5 Code setting criteria for a P0121.

MIL FUNCTION

The MIL is designed to indicate to the vehicle owner that something is wrong. Additionally, it is generally used by emission testing as an indication of a fail. Usually the MIL must be commanded on for the fail to occur as opposed to just being observed. There is a PID for MIL status that can be observed on a scanner. Remember that a PID or parameter identification is a single line of data that can be seen on a scanner.

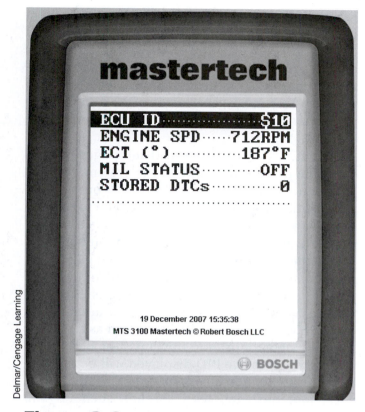

ECU ID·················$10
ENGINE SPD·····712RPM
ECT (°)··········187°F
MIL STATUS········OFF
STORED DTCs·········0

19 December 2007 15:35:38
MTS 3100 Mastertech © Robert Bosch LLC

BOSCH

Delmar/Cengage Learning

Figure 3-6 A scanner will show MIL status as either off or on.

MIL command status is used to fail or pass a vehicle during an emission test even if the actual MIL in the dash is burned out. You can see from Figure 3-6 that the status of the MIL on this vehicle is off.

If the MIL is commanded on, it indicates a few important points. First, a monitor has run and found a problem. Second, there should be freeze frame data stored within the PCM that can be useful. Freeze frame is a snapshot of the conditions present when the DTC is set. On the other hand, if the MIL is not on, it doesn't necessarily indicate the opposite conditions. If the MIL is not on, there still may be a problem unless it comes after all monitors have run to completion. If, for instance, a vehicle has generated the data shown in Figure 3-7, then the P0442 indicates that the EVAP (evaporative emission) monitor has run and has found a small leak.

You will see later on that the EVAP monitor is typically one of the more difficult monitors to run. It

has complicated enabling criteria. With the P0442, there should be an illuminated MIL and a set of conditions that were present when it set freeze frame. Let's follow this example and see how we should *not* proceed. When we were dealing with OBD I, it was common to clear the code to see if it came back. With OBD II, if we clear the P0442, we erase the DTC and the freeze frame data. Additionally, and more important, we clear the monitors and turn the MIL off (command status). Think about what it will take to have the MIL come back on. The key to this function are the monitors, which need to run to completion. If the enabling criteria cannot be easily met, it could take weeks or even months to get the monitors to run and reset the P0442 DTC. So, does "clear the code and see if it comes back" make sense with OBD II? The answer is no. When the MIL is on, the technician needs to scan the vehicle for all DTCs, freeze frame data, and monitor status. You will see in the future that this makes the most sense and will most likely yield a successful repair.

DTC TYPES

This is probably a good time to identify the two different types of DTCs available within the OBD II system. They are not difficult to comprehend and are based on how the monitor runs that will identify them. The simplest way to understand them is to realize their function. Let's start with a simple component DTC (Figure 3-8).

This DTC involves an out of range sensor. For this DTC to be captured, the monitor will run at least twice. The first one will identify the conditions that are present to set the DTC and will result in a pending DTC being set. This "temporary" or pending DTC will reside within the OBD system until the monitor runs again. If the condition that was identified is still present, the DTC becomes a hard fault, turning on the MIL, capturing freeze frame, and storing the code. This type of failure is frequently referred to as a *Type B*. Type B DTCs usually are not extremely important, and normally will not result in additional component failure. If, after setting a pending DTC, the monitor does not see the same issue the next time it runs, the pending DTC will be erased.

The other type of DTC is the *Type A*. It involves the possibility that additional damage might occur. As

P0442	Evaporative Emission System Leak Detected (small leak)

Figure 3-7 OBD II can test for small (0.20") leaks during the EVAP monitor. Delmar/Cengage Learning

P0121	Throttle/Pedal Position Sensor A Circuit Range/Performance

Figure 3-8 A DTC that resulted from a comprehensive component monitor. Delmar/Cengage Learning

P0300	Random Misfire Detected

Figure 3-9 This Type A DTC can result in CAT damage. Delmar/Cengage Learning

a result it will not be run twice, but will set the DTC after only running the monitor once. If, for example, the monitor sees a misfire problem, it might set what is shown in Figure 3-9.

The misfire monitor has seen sufficient misfire to determine that the MIL needs to be turned on, a freeze frame captured, and the DTC set. The difference between these two DTCs is the fact that, with Type A, there was no pending DTC set. Instead, the PCM went directly to the setting phase. Think about the reason why: there is a possibility that the misfire will allow additional fuel into the catalytic convertor, causing damage, so the pending phase was skipped and instead, the PCM went directly to the P0300 DTC. If the code setting criteria involve possible damage to some other component, especially the catalytic convertor, then the pending code phase is skipped. In theory, this should allow the consumer to see the MIL on and get the vehicle to you for diagnosis and repair. Don't forget that the DTC set could be within the generic or the manufacturer's side of OBD II.

PIDS

A PID is parameter identification, where "parameter" refers to a single line of data shown on a scanner.

For instance, engine speed is a PID; throttle position is another. All components that generate data will have PIDs. These PIDs will be displayed on a scanner and be a part of either the generic or the manufacturer's side of OBD II. Remember that State emission tests only look at the generic side of the system; there is far more data available on the manufacturer's side (Figure 3-10).

The figure shows one of three pages of data available on the OEM side of the system. It is not uncommon to see in excess of 400 PIDs on the OEM side. Additionally, it is important for you to remember that the generic side of the system is transmitting data at 1996 speeds, while the manufacturer's side of the system has steadily increased data transmission speeds just about every year. You will see why this is important when we discuss diagnosis using a scanner. As a technician looks at more and more PIDs, the system will slow down as it packets the data together prior to transmission. Simplified, this means that one PID displayed will be the fastest refresh rate. Add an additional PID and the data speed for refresh doubles. Each subsequent PID will decrease the speed of the scanner's display. Sometimes this slowing down will not have an impact on what you are doing, but when you decide to display the data in graph mode, speed becomes a serious issue. Figure 3-11 shows a scanner with only two PIDs on the screen. This data update rate will be just about as fast as it can be. This topic will be discussed in more detail later.

Available GM Scan Data 1

TECHVIEW

File Tester View mode Setup Window Help

Real-time Tester data

Engine Speed =	635 RPM		Spark Advance =	19 °
IAC =	62		TPS (V) =	0.54 V
TPS (%) =	0 %		Vehicle Speed =	0 MPH
ECT (°) =	86 °C		IAT (°) =	27 °C
Loop Status =	Closed		Engine Load =	3 %
MAP (P) =	32 KPa		MAF (R) =	4.7 g/s
BARO (P) =	99 KPa		Ignition (V) =	14.1 V
Desired IAC =	62		Desired Idle =	625 RPM
Rich/Lean Stat 1 =	Lean		Rich/Lean Stat 2 =	Lean
Long Term FT B1 =	134		Long Term FT B2 =	131
Short Term FT B1 =	127		Short Term FT B2 =	127
Air Fuel Ratio =	14.7		KS Active Cntr. =	133
EGR Duty Cycle =	0 %		EGR Position =	0 %
EGR Desired Pos =	0 %		Evap Purge D.C. =	17 %
Engine Run Time =	0:07:22		Startup ECT =	75 °C

Delmar/Cengage Learning

Figure 3-10 One-third of the data available on the OEM side of the PCM.

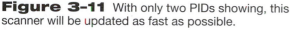

Delmar/Cengage Learning

Figure 3-11 With only two PIDs showing, this scanner will be updated as fast as possible.

CONCLUSION

In this chapter we looked at the DTC in detail. We decoded it to reveal what part of OBD II generated it, and the function of the monitoring system. We also looked at some of the forms that are used within State emission tests. The use of scanners relative to generic and manufacturers' data was also analyzed.

REVIEW QUESTIONS

1. Technician A states that the enabling criteria are the set of conditions required to get the monitor to run. Technician B states that a DTC can only be set if the monitor responsible for the DTC runs. Who is correct?

   a. Technician A only

   b. Technician B only

   c. Both Technician A and B

   d. Neither Technician A nor B

2. A catalyst efficiency is done by looking at what two signals?

   a. ECT and IAT

   b. MAP and MAF

   c. S1 O2S and S2 O2S

   d. Front O2S from both banks on a V-6

3. Technician A states that a sensor must pass the rationality test or a DTC will be set. Technician B states that a sensor must pass the functionality test or a DTC will be set. Who is correct?

   a. Technician A only

   b. Technician B only

   c. Both Technician A and B

   d. Neither Technician A nor B

4. A P0133 is set. Technician A states that this is a manufacturer-specific DTC. Technician B states that this is a generic DTC. Who is correct?

   a. Technician A only

   b. Technician B only

   c. Both Technician A and B

   d. Neither Technician A nor B

5. A generic DTC

   a. is the same for all manufacturers

   b. carries the P1 notation

   c. is only visible on a manufacturer's scanner

   d. is all of the above

6. Technician A states that the first part of an emission test is to scan for monitor status. Technician B states that MIL function and DTC printing will be done on a vehicle that has sufficient monitors ready. Who is correct?

   a. Technician A only

   b. Technician B only

   c. Both Technician A and B

   d. Neither Technician A nor B

7. Code setting criteria are the

   a. specific monitor used to generate the DTC

   b. set of conditions to run the monitor

   c. set of conditions to set the DTC

   d. rationality checking of a component

8. The function of the MIL is to allow

   a. the vehicle driver to know that something is wrong

   b. the State to pass/fail the vehicle for emission purposes

   c. a technician to recognize that a DTC is present

   d. all of the above

9. Technician A states that some of the monitors will test a complete system. Technician B states that some monitors will check down to the component level. Who is correct?

   a. Technician A only

   b. Technician B only

   c. Both Technician A and B

   d. Neither Technician A nor B

10. A PID is a

    a. single line of data viewed on a scanner

    b. part of a monitor sequence

    c. all of the freeze frame data

    d. only a manufacturer's specific data parameter

READINESS AND MONITORS

OBJECTIVES

At the conclusion of this chapter you should be able to: ■ Explain the purpose of a monitor ■ Look up enabling criteria and apply them to a vehicle ■ Analyze the enabling criteria looking for reasons why specific monitors will not run ■ Analyze a trip's enabling criteria ■ Analyze a drive cycle's enabling criteria ■ Analyze why monitors have not run

INTRODUCTION

In previous chapters, we discussed the basics of monitors. In this chapter, we'll look at the monitors in greater depth and will explore enabling criteria and drive cycle. We'll also spend extensive time analyzing the reasons that monitors might not run and will look in great detail at how the entire system functions. We'll use actual vehicle examples to help you understand this extremely important topic.

THE PURPOSE OF THE MONITOR

You will remember that the purpose of the monitor is to check a component or system. You will also recall that the only way that the MIL can be turned on and a DTC set is to run a monitor. If the monitor will not run, it is impossible for the system to generate diagnostic data. Make it a priority to always check monitor status as a starting point for diagnosis and repair. You will be tempted to only check the reason why the MIL is on and fix the DTC. This in itself is not a bad idea but will not tell the entire story. Frequently, when a DTC is set, some additional monitors will change their status from run to not-run. If a monitor has not run, then the system has not been checked. The usual sequence is that the customer has been driving around with the MIL off and all monitors run. So far, so good—until a monitor detects a problem, turns on the MIL, sets a DTC, and captures freeze frame data. At this point the PCM will likely suspend some other monitors and drop their status from run to not-run. That is not a problem unless the customer drives around for months with the light on. During this time additional problems might have developed that will go undetected because the monitors are not running. When the vehicle is finally brought in for service, the technician decides to fix the existing DTC. Once that is repaired, the other suspended monitors begin to run and pick up the additional problems, and the light (MIL) turns on again. If you put on your customer hat, you might think that the same problem has returned! "That technician"—actually, you—"must not have repaired my vehicle because the same problem is back." Remember, as far as the customer is concerned, the light only means one thing: the problem is back. You need to know that there are thousands of DTCs possible, and your customer will have difficulty understanding this concept.

The moral of this story is: always check the status of the monitors. If some are not run, explain to your customer that, after you repair the DTC, you need to get the monitors to run so the system will begin to check itself and might find and indicate an additional problem or problems. Don't forget the key to the OBD II system is an understanding of the monitors, enabling criteria, and drive cycles.

ENABLING CRITERIA

As we have mentioned, the key to understanding the monitoring system is the enabling criteria. The enabling criteria are the set of conditions necessary to get the monitors to run. If the enabling criteria are not met, the monitors will not run. The diagnostic system is only functional if the monitors will run. It is important to note that there really are no enabling criteria for the continuous monitors: misfire, comprehensive component, and fuel. These monitors should run almost immediately with very little driving. The issue is with the non-continuous monitors and their need to have very specific enabling criteria. Let's look at some case studies involving monitors that would not run.

Figure 4-1 shows the test results for a 1997 Mazda Protegé. The owner has been trying for months to get the required emission test accomplished for this vehicle. The customer had driven it under all types of conditions for an extended period of time. When the vehicle arrives at the service center, the customer is not happy and the State wants an emission test completed. The results of the latest emission test are shown in the lower section of the illustration. Notice that five monitors are listed as "NOT READY" and three are "Not Supported." "Not supported" is another way of saying these monitors are not on this vehicle and are of no concern. Basically, none of the applicable monitors have run. You will see later that 70% of the time this involves the specific enabling criteria for the vehicle. About 30% of the time something is wrong that is part of the enabling criteria.

Figure 4-2 shows part of the enabling criteria for all monitors. This is pretty straightforward, involving temperature, load, speed, and fuel criteria. The shop begins the diagnosis and sees that the majority of the non-continuous monitors have not run. A check of the applicable PIDs that involve the enabling criteria shows that everything seems to be present. There does not appear to be any reason why the monitors will not run if the vehicle is driven under the correct conditions. The technician takes the vehicle out and tries for over two hours to get the monitors to run. He follows as closely as possible the criteria that he has used previously on vehicles, with still no monitors run. At this point a call was made to the EPA to see if there were additional problems that they were aware of. The person at the EPA decided to go over the enabling criteria with the technician and asked if condition #3 had been met. Were all the electrical loads off? The technician indicates that everything is off except the radio, which is not much of an electrical load.

The key to understanding this case study is the ability to read between the lines with a lot of interpretation. In this case, simply turning the radio off allowed the monitors to run to completion. Apparently the vehicle's programming required minimal electrical loads running, and the radio was enough of a load to shut down the monitor system. The consumer, like many of us, had the radio on all of the time, and this prevented the system from running the monitors to completion. Once the monitors ran, no DTCs were set,

VIN		Plate		VIR **Z117740372**		Enf **SOS ENF**		**P**
Obs		XLck **N**	HExt **Y**	Vol **N**	OBD-ONLY **N**			
Year **1998**	Make **MAZD**	Model **PROTEG**	Cyl **4**	Type **P** Adap		Asn **12/2004**		
Status	**SELECTED FOR SOS ENFORCEMENT ON 04/09/2005**			Veh#		Exp **03/2005**		
Own 1				DLN		Reg **09/2005**		
Own 2				DLN				
Addrs				Zip		Cty **103**		
Fit **N**	OC **4** Des **002** Pur **10/02/1998**		Upd **10/19/2004**	Ta **12**	Clr **No Date**	**U**		
Pending suspension on:	**06/20/2005**							

OBD TEST RESULT - REJECT

We were unable to complete testing your vehicle today because the check of OBD readiness monitors indicates that the vehicle has not completed all required emissions component evaluations. The results of the OBD systems readiness check are listed below. Systems/components preventing the completion of the OBD test are listed as NOT READY. This can happen if you have recently had repairs performed on your vehicle, replaced the battery, or if the battery has run down or been disconnected.

Your vehicle did not fail the OBD test and this reject does not mean that there is anything wrong with your vehicle.

In order to prepare most vehicles for the OBD test, the vehicle should be driven for several days under a variety of normal operating conditions. This includes a mix of highway and stop and go, city type driving.

Your vehicle owner's manual may provide more specific instructions on preparing the vehicle for the OBD test. Additional information on preparing the vehicle for testing can be provided by consulting your dealer service department or independent repair facility.

CODES	DESCRIPTION	STATUS
CAT	Catalyst Efficiency Status:	NOT READY
CATHEAT	Catalyst Heating System Status:	Not Supported
EVAP	Evaporative System Status;	NOT READY
AIR	Secondary Air System Status:	Not Supported
AC	Air Conditioning Refrigerant Status:	Not Supported
02S	Oxygen Sensor System Status:	NOT READY
OS2HEAT	Heated Oxygen Sensor System Status:	NOT READY
EGR	Exhaust Gas Recirculation System Status:	NOT READY

Delmar/Cengage Learning

Figure 4-1 This vehicle was rejected for lack of monitor ready.

OBD II Drive Cycles	Mazda
	Protegé

All Monitors		**1998**
		All Engines

Notes	1. Disconnecting the battery will reset the PCM memory. 2. Monitor vehicle speed using the scan tool.
Conditions	1. Verify that ignition timing and idle speed are correct. If not, make necessary adjustments. 2. Verify that terminals TEN and GND of the DLC ore nor connected. 3. Verify that all electrical loads (headlights, blower, A/C, etc.) are OFF.
Step 1	Connect scan tool.
Step 2	Start the engine and warm it completely to operating temperature.
Step 3	Check the status of readiness monitors. Using the specific PID Monitor function, select "RFC FLAG" and press Start. If the "RFC FLAG" indicates YES, check the individual monitors. If the "RFC FLAG" indicates NO, go to Step 4.
Step 4	Run the engine at no load at 2300–2700 rpm for more than 15 seconds.
Step 5	Increase speed to 3800–4200 rpm for more than 15 seconds.
Step 6	Idle the engine for more than 20 seconds with the cooling fan stopped.
Step 7	Check the status of the monitors. Check for DTCs.

Figure 4-2 Enabling criteria for all monitors.

Vehicle Info and Test History

VIN		Plate		VIR **Z117740372**	Enf **SOS ENF**	**P**

VIN   Plate   VIR **Z117740372**   Enf **SOS ENF**   **P**
Obs   XLck **N**   HExt **Y**   Vol **N**   OBD-ONLY **N**
Year **1998**   Make **MAZD**   Model **PROTEG**   Cyl **4**   Type **P**   Adap   Asn **12/2004**
Status **SELECTED FOR SOS ENFORCEMENT ON 04/09/2005**   Veh#   Exp **03/2005**
Own 1   DLN   Reg **09/2005**
Own 2   DLN
Addrs   Zip   Cty **103**
Flt **N**   OC **4**   Des **002**   Pur **10/02/1998**   Upd **10/19/2004**   Ta **12**   Clr **No Date**   **U**
Pending suspension on: **06/20/2005**

Action	Vir#	Stn	Ln	Date	Time	Test Type	Test No.	Test Mode	Test Rslt	E C O	Rsn
Test	V012491109	17	2	10-11-2005	09:57	OBD	1	O	Pass	- P P	
Test	V012454390	17	2	06-23-2005	10:43	OBD	1	O	Reject	- P R	OM
Test	V012493608	17	2	06-13-2005	14:31	OBD	1	O	Reject	- P R	OM
Test	V012453923	17	1	06-09-2005	10:42	OBD	1	O	Reject	- P R	OM
Test	V013407721	16	3	06-01-2005	14:51	OBD	1	O	Reject	- P R	OM
Hold	Z120787112	36	38	05-04-2005	09:37						I
Test	V013395982	16	3	03-14-2005	12:36	OBD	1	O	Reject	- P R	OM
Test	V011155784	16	2	01-24-2005	11:16	OBD	1	O	Reject	- P R	OM
Test	V012505029	17	1	01-10-2005	12:49	OBD	1	O	Reject	- P R	OM

Figure 4-3 A pass after nine months of testing that simply required the radio to be off.

and the vehicle was taken in for its semiannual emission test.

A check of the State recording system (Figure 4-3) shows that the vehicle was originally taken in for an emission test in January 2005 (the bottom line of the Vehicle Info and Test History). At this time it was rejected, and the customer was probably told to drive the vehicle more. For the next nine months the consumer continued to bring the vehicle in and get it rejected. Look up toward the top of the form where the Status line is. This vehicle was "selected for SOS enforcement on 04/09/2005." SOS is the Secretary of State. The vehicle was required to have the test, and when it was rejected, the Secretary of State put it on the illegal list, effectively removing the license plates and registration. An unregistered vehicle cannot be insured, a situation made worse in an accident.

Getting this vehicle to run the monitors became as important for the repair shop and the technician as for the owner, and all it took was turning the radio off!

Reading between the lines on the enabling criteria is a very important part of getting monitors to run.

Figure 4-4 shows the enabling criteria from a Ford vehicle that had been rejected on numerous occasions. When the technician got the vehicle, the repair ticket stated the customer wants the vehicle to pass the emission test. A quick check of the appropriate PIDs indicated all was well. So the technician attempted to run the drive cycle using the enabling criteria and noticed immediately that the vehicle's speedometer did not work. Steps 3, 4, 6, 8, and 10 all require that the PCM know the vehicle speed. Apparently the customer had brought the vehicle to another repair shop but would not allow them to fix the speedometer. The other shop failed to note that the monitors would not run unless specific speeds could be met, so the customer refused the repair. The second shop indicated why the speedometer was important to the emission test and that without it, the vehicle would never pass. The customer finally agreed to the speedometer and the monitors set during the drive cycle using the enabling criteria. Reading between the lines to understand what the PCM is looking for is very important.

Let's read between the lines on another example from Figure 4-5.

What does the pre-conditioning information really mean? The first condition, "MIL must be off," is

| Ford, Lincoln, Mercury | OBD II Drive |
| All except Escort, Probe, Tracer, Aspire, Villager | Cycles |

| 1996 | All Monitors |
| All Engines | |

Notes	1. The primary intent of the Drive Cycle is to clear the P1000 code. 2. Rough road surfaces and high temperatures may prevent some monitors from completing. 3. For monitors other than HOS2, EGR and EVAP, run entire drive cycle.
Conditions	1. Disengage PTO if equipped. 2. Warm engine to at least 180 degrees F. 3. Do not turn engine OFF during drive cycle or interrupt the drive cycle. If so, repeat drive cycle.
Step 1	Start the engine. Drive or idle in N for 4 minutes.
Step 2	Idle vehicle in D (N on M/T) for 40 seconds.
Step 3	Accelerate to 45 mph (M/T shift from 1st to 5th, but hold 2nd for 5 seconds) at 1/4–1/2 throttle for 10 seconds.
Step 4	Drive at steady throttle at 45 mph (M/T use 5th gear) for 30 seconds.
Step 5	Idle the vehicle in D (N for M/T) for 40 seconds.
Step 6	Drive in city traffic between 25–40 mph (M/T in 3rd or 4th where possible) for 15 minutes.
Step 7	Step 6 must contain at least 5 stop and idle modes for 10 seconds each; 5 accelerations from 1/4–1/2 throttle; 3–1.5 minute steady state drives at 3 different speeds.
Step 8	Accelerate to between 45–60 mph (up shift to 5th gear in M/T). This should take about 8 minutes.
Step 9	Drive vehicle at steady throttle at selected speed (Step 8) for 5 minutes.
Step 10	Drive vehicle for 5 minutes at varying speeds between 45–60 mph (M/T use 5th gear).
Step 11	Bring vehicle back to idle. Idle in Drive (N for M/T) for 40 seconds.
Step 12	Check status of monitors.

Delmar/Cengage Learning

Figure 4-4 All monitor enabling criteria.

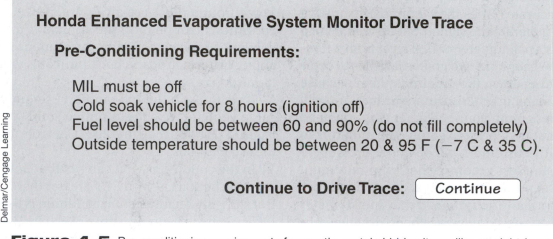

Figure 4-5 Pre-conditioning requirements frequently contain hidden items like an eight-hour cold soak.

important because it indicates that this monitor will not run if there is an active DTC and the MIL is on. You will have to clear the DTC and turn off the MIL prior to trying to get the monitors to run. As we get deeper into monitors, you will see that we typically do not want to clear DTCs that were fired off of the continuous monitors because that clears all monitors and resets them to not-ready. However, to get this non-continuous monitor to run, the MIL must be off. What this means is that as a technician, you will have to distinguish between DTCs that were fired off of the continuous monitors and those that the non-continuous monitors were involved in. After the repair is complete, you will have to clear the DTC to get the monitor to run. Make sure that you have copied down all information held in the PCM's memory prior to clearing the DTC(s).

The second condition is perhaps the most difficult one to both comprehend and test for. "Cold soak vehicle for 8 hours (ignition off)" is not exactly what it seems to be. On the surface it appears that we just have to let the vehicle sit overnight in the shop to satisfy the eight-hour key-off condition. However, reading between the lines, we ask ourselves: Does the PCM know that the key has been off for eight hours? Is there a clock circuit that knows what time it is? The answer to both of these questions is no. The PCM does not have a clock within itself. Then how does it know that an eight-hour soak occurred? This is the reading between the lines part; the PCM will compare the temperatures of the engine using the ECT and the intake (ambient) air using the IAT. A cold soak has occurred when the temperatures are the same, when engine temperature and air temperature are equal. Indirectly this also means that these two sensors

must be accurate within a couple of degrees. Generally, manufacturers will use a temperature of around 10 degrees as the maximum difference between IAT and ECT to indicate a cold soak has occurred. But what if the ECT is off by 8 degrees reading high and the IAT is off by 5 degrees reading low? When the actual ECT and IAT are at 50 degrees, the ECT will see 58 degrees Fahrenheit, and the IAT will see 45 degrees Fahrenheit. This is a difference of 13 degrees, with the PCM looking for less than 10. Will the EVAP monitor ever run on this Honda? The answer is no. A condition to running this EVAP monitor cannot be met and will therefore not run. Make sure that you check the accuracy of the ECT and IAT when you see an eight-hour soak as a requirement to running a monitor.

The last two conditions are pretty straightforward. The gas tank cannot be over or under filled. Additionally, this means that the gas gauge must work, or the PCM has no way of knowing the amount of gas in the tank. The last condition of outside temperature being between 20 and 95 degrees F at startup is going to be checked by the IAT, so again the accuracy of the sensor comes into question. The minimum of 20 degrees is pretty reasonable, but some Ford products have a minimum of 45 degrees Fahrenheit. This is not a problem in the South, but think about the northern states. It is likely that the EVAP monitor will not run during December, January, or February. As soon as warmer temperatures of spring arrive, the monitor runs and detects problems that occurred during the three months that it did not run.

The ability to read between the lines when looking at the enabling criteria is extremely important. Ask yourself how does the PCM know that the condition has been met? Do the enabling criteria require a signal from a sensor or sensors that will need to be accurate? Does the required sensor have the ability to generate the signal? Speed read off of a nonfunctional speedometer or fuel level from a nonfunctional fuel gauge are two good examples of components shutting down the monitor system because a part of the enabling criteria cannot be met. You will need to impress upon your customer why you need to fix the component prior to getting the monitors to run. Frequently customers will be under pressure to get the emission test accomplished with minimum expense to keep the vehicle legal. They might believe that they really do not need that speedometer or gas gauge. After all, they have been driving without it for months without problems. The pressure to get the test accomplished becomes the number one priority. Your job will be to tactfully explain the interrelationship and the necessity of the missing input to the monitor. When dealing with customers or even other technicians it is important to keep in mind the basic function of the system as an emission testing strategy. If you understand that the basic function involves testing, and that testing is another way of saying running enabling criteria so that monitors will run, then the importance of the monitors will not be understated.

DRIVE CYCLE

A few terms and phrases are important for your overall understanding of OBD II. The first is monitor, which by now you realize is a test done on a system that requires enabling criteria to run. The enabling criteria are the set of conditions required by the PCM before it will test the individual systems. When the enabling criteria have been met for one monitor, a *trip* has occurred. The last phase is the drive cycle. The drive cycle is the set of enabling criteria that will allow all of the monitors to run. If all trips are run the same, results will be found while running the drive cycle. When you refer to maintenance literature, you might be given the choice of setting an individual monitor (a trip) or setting all monitors (a drive cycle). Technicians need to have scanned the vehicle to identify which monitors need to be run. If you are faced with only one monitor not ready, then it makes sense to drive the enabling criteria for just the one monitor.

Figure 4–6 shows the catalyst monitor for a 2004 Chevrolet Aveo. The pre-condition section of the enabling criteria indicates that the vehicle must be warmed to at least 158 degrees F but not hotter than 228 degrees F and must be in closed loop. Intake air temperature is also indicated, plus minimum air flow through the mass air flow sensor. Once these conditions can be met, the vehicle is deliberately driven at highway speeds. After meeting the enabling criteria, a scan check indicates whether the monitor has run.

If we look up the "all monitor" sequence on the same vehicle, we should notice some differences. Figure 4–7 shows the same 2004 Chevrolet Aveo under these conditions.

In the conditions, you can see that now the vehicle must be started with ECT greater than 23 degrees F at start-up. An interesting condition is listed for the IAT. IAT at start-up must be less than 104 degrees, and as soon as the engine starts must be within 5 degrees of the key on–engine off (KOEO) temperature IAT. This is just another way of indicating that the engine is truly completely cooled off. KOEO temperature is read immediately prior to start-up. Once the engine is running, it will pull in true ambient air, which must be the same as that of the engine. Once these conditions have been met, the vehicle is driven specifically at low speeds and high speeds until all monitors set. All monitors will be set by driving the drive cycle.

Here is where the importance of noting in advance just what monitors have run should be noted. If the above Aveo comes into the shop without the CAT monitor run, after working on the vehicle, the technician can satisfy the enabling criteria assuming no one has cleared the codes. Clearing the codes resets all monitors to a not-run status, which will require a cold soak probably lasting overnight. If you pay attention to the monitor status during the diagnosis of the vehicle, you will know in advance if it needs to be kept overnight or can be fixed, emission tested, and given back to the customer the same day. The other very important thing to note is how complicated the all monitors procedure is. It has very specific conditions that must be met. In certain areas of the country, notably in cities where traffic is an issue, the technician might have to find an area with little-used streets to be able to satisfy the enabling criteria.

WHY MONITORS WILL NOT RUN

It is important to emphasize that the heart of the modern OBD II system are the monitors. They must run to completion or the system is nonfunctional. In states where there is mandatory emission testing, and vehicles are rejected for lack of monitors run, a 10 to 20% reject rate is common. In large states like

OBD II Drive Cycles

General Motors
Chevrolet Aveo

Catalyst Monitor	2004
	1.6L

Notes	1. The engine must run for at least 10 minutes to initiate this test. 2. The A/C compressor must not be cycling.
Conditions	1. ECT 158–228 degrees F; engine in closed loop. 2. IAT 19–221 degrees F. 3. Air flow into engine >12 g/s for 45 seconds. Vehicle must be driven, then allowed to idle.
Step 1	Perform the I/M System Check. Failure to do so may result in difficulty in updating the monitor (s) status to YES.
Step 2	Ensure all conditions have been met. Turn OFF all accessories.
Step 3	Start the engine and allow it to idle.
Step 4	Accelerate at part throttle to 55 mph and maintain speed for 2 minutes.
Step 5	Decelerate to 0 mph. Idle for 2 minutes or until I/M System Status updates to Yes.
Step 6	If the Catalyst monitor is YES, go to Step 14.
Step 7	Using a scan tool, access the DTC information. If there are any failed DTCs, diagnose and repair the DTCs.
Step 8	Determine which DTCs are required run in order to complete the test.
Step 9	Using a scan tool, observe the Not Ran Since Code Cleared display. Determine which DTCs that are required to have run, and have not run, for a YES status to be displayed.
Step 10	Enter the DTC in the specific DTC menu of the scan tool.
Step 11	Operate the vehicle within the Conditions Required for Running Monitors. Repeat the procedure until the scan tool indicated the diagnostic test has run.
Step 12	Repeat Steps 9–10 for any DTCs that have not run.
Step 13	Check the I/M System Readiness display. If the AIR system status is YES, go to Step 14. If not, refer to service manual.
Step 14	Check for emission-related DTCs. If any are present, go back to Step 7. If no DTCs are present, AIR system status is OK.
Step 15	

Delmar/Cengage Learning

Figure 4-6 Running a single monitor is called "a trip".

General Motors	OBD II Drive
Chevrolet Aveo	Cycles

2004   **All Monitors – Complete System Set Procedure**

1.6L

Notes	1. Use this drive cycle when 2 or more or all readiness monitors are set to NO. 2. Preprogramming the scan tool will shorten test length.
Conditions	1. BARO > 72kPa; ECT > 23 degrees F at start-up 2. IAT < 104 degrees F; Start-up IAT minus actual IAT within 5 degrees F. 3. Ignition 1 voltage 10–16; fuel level 25–75%
Step 1	The engine must be OFF for more than 6 hours. Start-up IAT minus start-up ECT within 22 degrees F. Start-up ECT minus start-up IAT within Perform the I/M System Check. Failure to do so may result in difficulty in updating the monitor (s) status to YES.
Step 2	Turn ignition ON, verify enabling conditions, the turn ignition OFF for 5 minutes. Preprogram the scan tool with vehicle information before turning ignition ON. Start the engine and do not turn it OFF for the remainder of the test.
Step 3	Turn OFF all accessories and set parking brake. A/T should be in P and M/T should be in N. Idle the engine for 5 minutes.
Step 4	Accelerate at part throttle to 45–50 mph and maintain speed until engine reaches operating temperature. This may take 8–10 minutes. Continue operating for 6 minutes.
Step 5	Accelerate at part throttle to 55 mph and maintain speed for 2 minutes.
Step 6	Decelerate with throttle closed for more than 10 seconds to 0 mph.
Step 7	Idle for 2 minutes with the brake pedal depressed and automatic in D.
Step 8	Accelerate to 55 mph and maintain speed fro 2 minutes.
Step 9	Release the throttle and allow the vehicle to decelerate to 20 mph. Repeat as necessary until the EGR monitor sets.
Step 10	Accelerate to 55 mph and maintain speed for 2 minutes.
Step 11	Decelerate to 0 mph and idle for 2 minutes.
Step 12	Access the readiness status on the scan tool. Perform the individual drive cycle for any monitor that does not display YES (monitor is not ready).
Step 13	Check for DTCs. Any DTCs will require diagnosis and repair.
Step 14	Following repairs and clearing DTCs, perform Steps 1–10 again or perform steps for individual monitors that are not set.
Step 15	

Delmar/Cengage Learning

Figure 4-7 All monitors should run during a drive cycle.

Readiness/Rejects

10.9%
1st Retest Reject Rate

2 Possible Reasons:

• Drive Cycles after repair

Approx. 70%

• Something preventing
monitors from running

Approx. 30%

Delmar/Cengage Learning

Figure 4-8 Rejects occur for a
variety of reasons.

where close to 2 million cars may be tested each year,
this equates to 100,000 to 200,000 vehicles that will
be rejected annually because insufficient monitors
have run. Your job as a technician will be to understand that 70% of the time, insufficient monitors run,
indicating that the specific drive cycle has not been
accomplished during the customer's normal daily
driving. The remaining 30% of the vehicles have
something wrong that is preventing the monitors
from running (Figure 4–8).

The list of possible causes is long and includes
some easy items such as sensors out of range or signals that are missing. You will see in future chapters that we have to use many different tools and
techniques to identify the cause of the monitors not
running.

CONCLUSION

In this chapter, we examined in detail the enabling
criteria and how they relate to lack of monitors. Using
some examples, we looked at the reasons why certain monitors may not have run and talked about the
specific reasons that a certain percentage of vehicles
will not run monitors. We also looked at some pre-
conditioning requirements that can cause monitors
to run.

REVIEW QUESTIONS

1. Technician A states that the purpose of the monitor is to help the technician repair the vehicle.
Technician B states that monitors will not run
unless enabling criteria have been met. Who is
correct?

a. Technician A only

b. Technician B only

c. Both Technician A and B

d. Neither Technician A nor B

2. A vehicle is repaired, DTC cleared, and driven
back for an emission test. The vehicle is rejected.
Why?

a. The vehicle is really not repaired yet.

b. The State needs to see the vehicle with the
DTC still present.

c. The monitors have not tested the system yet.

d. The check engine light is off.

3. Two technicians are discussing enabling criteria.
Technician A states that the enabling criteria will
always be met by driving under normal conditions. Technician B states that the scanner can
be used to set all monitors while still in the shop.
Who is correct?

a. Technician A only

b. Technician B only

c. Both Technician A and B

d. Neither Technician A nor B

4. Why is it important to run the monitors after a
completed repair?

a. The monitors will indicate, once they run, that
the vehicle is truly repaired.

b. The monitors will turn off the check engine light,
which is the only way it can be extinguished.

c. The State requires it.

d. The enabling criteria are different before and
after repair.

5. A vehicle has a bad thermostat (stuck open). This
might

a. cause a monitor to *not* run

b. cause engine temperature to be too low, especially in winter

c. result in a DTC being stored

d. all of the above

6. Technician A states that if the enabling criteria
call for a specific speed, reading between the
lines would indicate that the speedometer must
be functional. Technician B states that if fuel level
is part of the enabling criteria, a functioning fuel
gauge is required. Who is correct?

a. Technician A only

b. Technician B only

c. Both Technician A and B

d. Neither Technician A nor B

7. Two technicians are discussing pre-conditioning requirements. Technician A states that this is the part of the enabling criteria that will set a DTC. Technician B states that the monitor will not run until these individual items are present. Who is correct?

   a. Technician A only

   b. Technician B only

   c. Both Technician A and B

   d. Neither Technician A nor B

8. "MIL must be off" is seen in the pre-conditioning requirements. Technician A states that the MIL should be turned off or reset by using a scanner. Technician B states that the monitor will not run if the MIL is on. Who is correct?

   a. Technician A only

   b. Technician B only

   c. Both Technician A and B

   d. Neither Technician A nor B

9. Cold soak vehicle means

   a. let it sit for a couple of hours

   b. keep it in a heated garage overnight

   c. allow the engine temperature and intake temperature to normalize and equalize (ECT and IAT)

   d. run the vehicle until the cooling fan cycles a few times

10. Two technicians are discussing the drive cycle. Technician A states that the drive cycle will generally allow all monitors to run. Technician B states that all of the trips taken together equal the drive cycle. Who is correct?

    a. Technician A only

    b. Technician B only

    c. Both Technician A and B

    d. Neither Technician A nor B

MONITORS AND STRATEGIES

OBJECTIVES

At the conclusion of this chapter you should be able to: ■ Explain the differences between continuous and non-continuous monitors ■ Identify the enabling criteria for most continuous monitors ■ Use "drive the freeze frame" to clear the MIL ■ Analyze the difference of the three continuous monitors ■ Use fuel trim to verify a repair ■ Use enabling criteria to get a non-continuous monitor to run ■ List reject criteria for different year vehicles

INTRODUCTION

We have already discussed why the heart of the OBD II system is the monitoring system. We also looked at some of the enabling criteria required to run a monitor. It is now time to begin the analysis of the monitors in moderate detail. In future chapters we will look at each monitor, enabling criteria, problems, and repairs. In this chapter we present an overview of each available monitor.

CONTINUOUS VERSUS NON-CONTINUOUS MONITORS

First, we need to separate the various monitors into two different categories: continuous and non-continuous. Their name gives a hint as to how they work. The biggest difference is in the enabling criteria. The enabling criteria for a continuous monitor are generally to have the engine start and run, while the criteria for a non-continuous monitor involve various driving, load, temperature, fuel, and speed components. Generally, when you scan for continuous monitor status, you will find them in a run or completed status. The simple start and run enabling criteria are met almost immediately and are running over and over again as the vehicle drives around under normal conditions. Because the continuous monitors are almost always in a run or completed status, most states do not scan them as part of their required emission test. The non-continuous monitors, those that require complicated enabling criteria, are frequently found in a not-completed state. For this reason states with

emission testing will scan the non-continuous monitors and either reject or fail a vehicle with insufficient completed monitors. If you scan for non-continuous monitors, it is common to have less than all of the available monitors run. Figure 5–1 shows the monitor status on a 2003 Lincoln.

The first three monitors; misfire, fuel sys (fuel systems), and comp (comprehensive component) show their status as COMPL (completed). Note: Some scanners will indicate the status as run. The remaining eight monitors (ten on some vehicles) show COMPL for completed, N/A for not applicable, or INCMPL for not-completed monitors. This is a very common scan

Figure 5–1 The monitor or readiness screen shows the current status of all monitors.

Delmar/Cengage Learning

showing most monitors completed and EVAP not completed. Simplified, this means that the enabling criteria for everything except EVAP have been met, and the various systems have been tested and either passed or failed. The monitor status screen does not usually indicate pass or fail and will not indicate DTCs. It is also important to note that having the EVAP monitor not run does not necessarily mean that the EVAP system has a problem. It just indicates that something prevented the EVAP monitor from running to completion. It might be that the enabling criteria contained some specific driving conditions that were too difficult to attain, or that some sensor required for the complete testing was not functional. The one, hopefully obvious conclusion is that the EVAP system might have something wrong but the lack of running the monitor prevents the system from identifying the problem. The only way the system can diagnose itself is to run monitors. If the monitor will not run, the system cannot and will not generate the DTC.

CONTINUOUS MONITORS

Let's start with the continuous monitors. There are three: misfire, fuel and comprehensive component. Generally these continuous monitors will be responsible for generating close to 50% of the DTCs that are common. The misfire monitor has the ability to generate, off of engine speed changes, information regarding power from each cylinder. It can generate a random misfire DTC (P0300) or get very specific down to the offending cylinder. A P0301 indicates that cylinder #1 is misfiring, while a P0302 indicates cylinder #2. These DTCs will not tell you why a specific cylinder is misfiring but at least give information that is helpful when diagnosing. Remember that there is really no enabling criterion for any of the continuous monitors, so it is possible that a PCM will detect a misfire at any speed/load combination. However, the misfire monitor generally requires that the engine speed be stable to run to completion. If the monitor runs and detects a misfire, it will set a P030X DTC, capture freeze frame, and turn on the MIL.

Let's look at freeze frame. At the moment that the PCM runs the monitor and detects a problem, a snapshot is taken of engine conditions. This snapshot is called freeze frame, and it is an indicator of just what the vehicle was doing around the time the DTC was set. Let's take an example of a misfire and look at freeze frame data for it. Figure 5–2 shows a scanner indicating that one DTC is present, a "P0302 Cylinder 2 Misfire Detected." Anytime you see a DTC, the first thing to do is try to identify which monitor was responsible for setting it. Once the DTC's number and short description are written down, look at freeze frame.

Figure 5–3 shows freeze frame data for this P0302. It is a snapshot of the conditions that were present at the time the DTC was set. Let's look at the data that it shows.

Delmar/Cengage Learning

Figure 5-2 This P0302 indicates that Cylinder 2 has misfired.

Parameter	Value	Units
Engine Speed	662	RPM
ECT	86	°C
Vehicle Spd	0	km/h
Engine Load	2.7	%
MAP (P)	30	kPaA
MAF (R)	5.1	gm/s
TPS (%)	0.0	%
Fuel Stat 1	CL	
Fuel Stat 2	CL	
ST FT 1	2.3	%
LT FT 1	4.7	%
ST FT 2	15.6	%
LT FT 2	9.3	%

Delmar/Cengage Learning

Figure 5-3 Freeze frame for a P0302.

Remember that this is a snapshot of conditions. In this case it appears the engine was fully warmed up (ECT = 85 degrees C), idling (662 RPM), and not moving (vehicle speed 0 km/h). The throttle is closed (TPS = 0.0%) with very little load (2.7%). It sounds like the vehicle was idling in park or neutral. There are two oxygen sensors up front. One for bank 1 (the bank that has the number one cylinder in it) and one for bank 2. These are labeled as Fuel Stat 1 and 2, and both show closed loop operation (CL). Closed loop indicates that the signal from the oxygen sensor (O2S) is being used to adjust the fuel delivery. The SFTF 1 and 2 indicate the amount of fuel trim being used to maintain fuel control. In this case bank 1 has a total correction of 7% (adding the 2.3% + 4.7% from 1) Bank 2 has a total correction of 24.9%. Both banks are adding fuel, with bank 2 adding much more than one. The P0302 indicates that cylinder #2 is missing, and you have probably guessed by this time that cylinder #2 is in bank 2, the one with the majority of the correction.

This next part is somewhat complicated, so don't hesitate to read it a couple of times until it makes sense. When a continuous monitor fires off a DTC, it sets in PCM memory freeze frame conditions. The conditions also become the enabling criteria for rerunning the monitor next time. So in our example of the P0302, if we fix the misfire, we can use the system to rerun the monitor by matching the conditions of freeze frame. In this example the P0302 occurred because of an open plug wire, and so we decide to install a set of wires. After the wires are installed we can take the vehicle out for a test drive and match the freeze frame conditions. By "driving the freeze frame," we satisfy the enabling criteria set in memory when the DTC was set. The monitor reruns twice, and if the misfire is gone, turns off the MIL. The DTC remains in memory, but the MIL is off. You might ask why you would want to do this. The answer is simple: if the continuous monitor reruns and turns off the MIL, then the vehicle is ready for an emission test. If, on the other hand, we clear the DTC using a scanner, we will reset all monitors to a not-complete status, requiring an extensive drive cycle prior to the emission test. The process of driving the freeze frame allowed us to easily rerun the misfire monitor and turn the MIL off after the repair without having to clear the DTC with a scanner. In this example, all we had to do was start the engine, let it idle until it was fully warmed up, and the MIL went out. Having the knowledge of driving the freeze frame easily saved us an hour. This was possible because we noted that the rest of the monitors had run prior to the repair, and we needed to just

get the MIL off without clearing the P0302. The ability to drive the freeze frame and clear the MIL only works for continuous monitors.

The next continuous monitor is the fuel systems monitor. This monitor directly looks at the two types of fuel trim; short term and long term. Fuel trim is the PCM's way of compensating for conditions that have changed over time or miles. For example, let's take a vehicle that has a fuel pump that has reduced volume. As the volume of the pump gets lower and lower, eventually it will have an impact on the amount of fuel injected. The system will begin to go lean, and the O2S signal will indicate this lean condition. The PCM will increase the pulse width of the injectors to compensate for the reduced fuel delivery. The increase will start out as an increase in short-term fuel trim (STFT) and will show as a +%. Once short-term fuel trim has the system back in fuel control, it will begin transferring the trim over to the long-term fuel trim (LTFT) with the goal of having 0% change in short-term fuel trim. STFT is stored in a volatile memory, and volatile memories reset when the key is turned off. Once the original correction is calculated by STFT, it is transferred to LTFT because it is stored in keep-alive memory circuitry. Keep-alive memory is not volatile and will be available as soon as the key is turned back on.

When we look at the results of fuel trim, we are looking at the results of the fuel system monitor running. The results are directly related to the signal off of the O2S, and because the monitor is running continuously, it is updated frequently. Figure 5–4 shows a vehicle that has a P0171 DTC (lean bank 1).

A P0171 DTC is an indication of a lean condition, and the scanner data shows that the fuel adjustment has been made and moved over to LTFT. Notice that STFT shows 0.0%. At the same time, LTFT is indicating a +14.8% of additional fuel being injected in an effort to compensate for a lean condition. Again, if we understand how the continuous monitors run, if we fix the vehicle and do not clear the codes, the system will begin running the monitor and make an adjustment to STFT. On the vehicle indicated here, a vacuum leak was the cause of the lean condition. After the repair of the leak, the engine was started. After a short test drive, again matching the conditions of the freeze frame, fuel trim began to compensate for what actually was now too much fuel. Figure 5–5 shows the result of driving the freeze frame.

Notice that the STFT is equal to the LTFT but is minus. So when we had the leak, fuel trim added fuel until the engine was in fuel control. After the repair, there was now too much fuel and STFT adjusted it down. The key to this situation is understanding that again it was

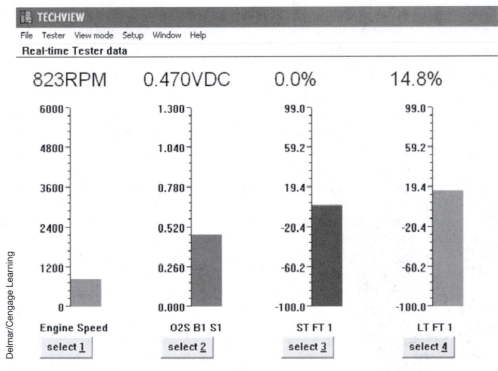

Figure 5-4 Trim as a result of a P0171 (lean) DTC.

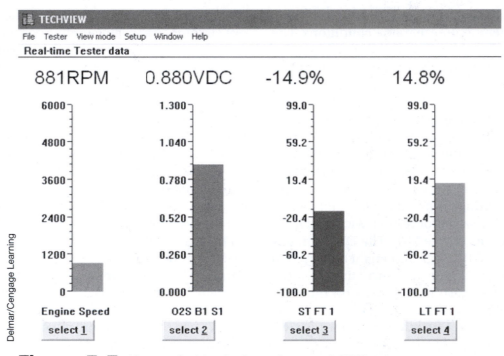

Figure 5-5 After repair, drive the freeze frame until STFT occurs.

not necessary to clear codes. Clearing codes would have reset all monitors and required a drive cycle. The drive cycle could have taken over an hour to accomplish, thereby delaying the emission test. Once the freeze frame was driven, the trim numbers indicated that the vehicle was repaired. An LTFT of +14.8% and an STFT of –14.9% cancel out one another, resulting in a trim of basically 0%. This vehicle is now fixed, and once the monitor reduces both trim values through a short test drive, the MIL will go out and the vehicle can be tested with no need to drive the drive cycle.

The last continuous monitor is the comprehensive component monitor. This monitor is most like OBD I. It will test circuitry and components like TP (throttle position), MAP (manifold absolute pressure), VSS (vehicle speed sensor), ECT (engine coolant

P0115 Engine Coolant Temperature Sensor 1 Circuit

Figure 5-6 A DTC set by the comprehensive component monitor. Delmar/Cengage Learning

temperature), etc. This monitor generally results in very clear and concise DTCs.

The P0115 in Figure 5-6 is a good example of the type of DTC generated by the comprehensive component monitor. It indicates that there is a problem within the circuit of the ECT. The monitor will have also turned on the MIL and again captured freeze frame data. Once the circuit is diagnosed and repaired, driving the freeze frame again should turn off the MIL, allowing the emission test to be accomplished.

To summarize the main points of the continuous monitors:

They run with virtually no enabling criteria.

They will support drive the freeze frame to turn the MIL off.

They result in DTCs that are frequently very descriptive.

They will be responsible for about half of the DTCs that you will see in the field.

State emission tests do not scan for their status. It is assumed that they have run.

It will not be necessary to clear the DTC and turn off the MIL to get them to run.

If you keep these points in mind, your experiences with continuous monitors and the DTCs that they generate will not be difficult.

NON-CONTINUOUS MONITORS

There are some major differences in dealing with non-continuous versus continuous monitors. The most important point is that they will run only when the enabling criteria have been met. The criteria will never be as simple as those for the continuous monitor: start and run. Instead, they will be more complicated and harder to attain. Also, driving the freeze frame to turn off the MIL is simply not practical. Look at Figure 5-7. It is from a Lexus vehicle that has a set of conditions that include "MIL must be off."

In theory we could drive around forever with the MIL on. The MIL cannot go off because it would require the running of the monitor, and the monitor will not run until the MIL is off. Frequently, the enabling criteria for non-continuous monitors will say "MIL must be off." Your only choice is to clear the DTC using a scanner before you attempt to satisfy the driving conditions of the enabling criteria.

Lexus All Models	OBD II Drive Cycles
2004 **All Engines**	**O2 Sensor or O2 A/F Sensor Heater Monitor**
Notes	1. Monitor status will be rest to "Incomplete" if the ECU loses power, DTCs are cleared or conditions for enabling conditions have not been met. 2. If a drive cycle is interrupted, it can be resumed. 3. Avoid sudden speed changes.
Conditions	1. MIL must be OFF.
Step 1	Connect a scan tool and check monitors status and precondition.
Step 2	Start the engine and idle for 500 seconds or more.
Step 3	Drive at 25 mph or more for at least 2 minutes.
Step 4	Check the monitor status. It should switch to "Complete". If not, be sure enabling criteria are met.
Step 5	If the monitor has not switched to complete, turn the ignition OFF and repeat Steps 2–3.

Delmar/Cengage Learning

Figure 5-7 Condition 1 means that the MIL must be off.

We will take an overview approach to each non-continuous monitor in this chapter. Later on, each monitor will be discussed in detail in its own chapter. Now let's look again at our scan from a Ford product (Figure 5–8).

The non-continuous monitors begin after the COMP MON (comprehensive component). The eight monitors that follow have the notation of COMP for completed, N/A for not applicable, or INCMPL for not completed. Let's start with the easiest: N/A. N/A shows for the HTD CAT MON (heated catalyst), the 2nd AIR MON (secondary air), and the A/C MON (air conditioning). Since the vehicle does not have these systems, the monitors are disabled and thus listed as N/A. Virtually no vehicle has a heated catalyst; and this vehicle does not have AIR (Air Injection Reaction), where additional air is injected into the exhaust stream. The last N/A monitor, the A/C MON, does not indicate that the vehicle doesn't have air conditioning, rather that it does not have R-12 (Freon) for refrigerant. This vehicle has R-134a, which is not a fluorocarbon and so does not have to be monitored. We have five monitors that should run, however. Don't forget the standard for most emission states for non-continuous monitors (Figure 5–9).

This scan is from a 2003 Lincoln, so we would be allowed to take the test with only the CAT monitor not complete. The first step in most State emission programs is to scan for MIL status. If the MIL is off, monitor status is scanned for the correct number based on the year of the vehicle. If the MIL is on, the vehicle fails with no need to scan monitor status. You might be wondering why the State does not scan for monitors if the MIL is on. The answer is simple. If the MIL is on, we know that at least one monitor ran, and it detected a problem, so the vehicle is failed.

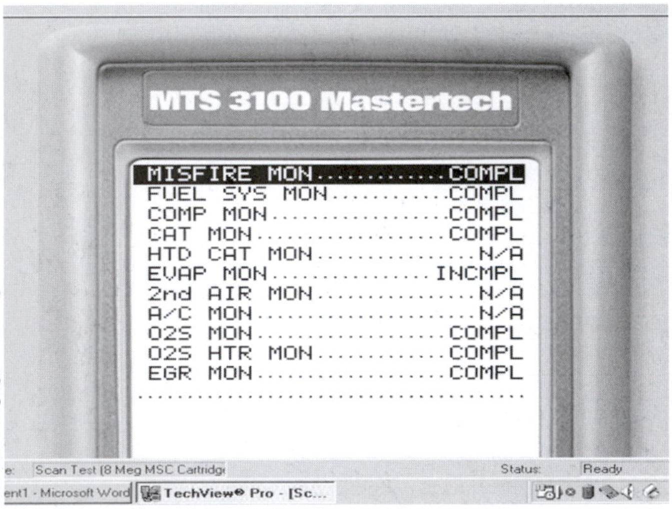

Figure 5-8 Readiness status indicates one non-continuous monitor incomplete.

Reject...

- 3 or more readiness flags
  NOT set (96-00)
- 2 or more readiness flags
  NOT set (01-newer)

Chart 3: Scan Tool Monitor Display

Continuos Monitoring Tests		
Monitor	**Availability**	**Status**
✓ Misfire	Supported	Complete
✓ Fuel System	Supported	Complete
✓ Component	Supported	Complete
Non Continuos Monitoring Tests		
Monitor	**Availability**	**Status**
✗ Catalyst	Supported	Not Complete
⊘ Heated Catalyst	Unsupported	
⊘ Evaporative System	Unsupported	
⊘ Secondary Air System	Unsupported	
⊘ A/C System	Unsupported	
✓ Oxygen Sensor	Supported	Complete
✓ Oxygen Sensor Heater	Supported	Complete
✓ EGR System	Supported	Complete

Figure 5-9 State inspection standards.

This brings up an interesting dilemma for some technicians. When the MIL is on, frequently some monitors are set to not-complete status and stay there because the MIL is on. Now the customer drives around for an extended period of time with the MIL on, and limited monitors run. The vehicle fails the emission test and is brought in for repairs. The technician fixes the problem and then runs the enabling criteria. During the time that the monitors were suspended (not completed), additional problems might have developed that went undetected. Now the monitors run and additional DTCs are fired off. The customer doesn't understand why the bill is more than the estimate, and the technician looks either incompetent or unethical—not a good impression to leave the customer with.

This problem can be resolved before it develops by maintaining control of the vehicle. With the MIL on, scan for monitor status. Write the not-completed monitors on the work ticket and explain to the customer that once the original DTC is fixed, the enabling criteria will be met and the other monitors will run. When they run, it is possible that additional DTCs will set that will have to be fixed prior to having the vehicle tested and passed. If the customer knows in advance that the repair sequence will consist of fixing the DTC, running the monitors, and then fixing additional DTCs prior to getting the emission test run, he is more likely to go along with the total repair.

The most important thing to keep in mind when dealing with non-continuous monitors is that, when they are running, they will follow a very specific sequence and test procedure under a variety of conditions. Take the first one, CAT (catalyst), for example. Most CAT monitors require sufficient running to fully warm up the catalyst and cannot see any DTCs (MIL on). Then they require some low-speed cruise operation, followed by usually above 45 mph for an extended distance. During the running of the monitor the PCM will look at O2S signals from before the CAT and signals from after the CAT. A comparison of the input and output signal frequency will indicate to the PCM the ability of the CAT to function. Chapter 11 on catalyst will go into detail as to how the signals are interpreted.

The next monitor to look at is the toughest one: EVAP. The evaporative emission system is a complicated system with numerous valves, lines, components, and electrical circuits. Manufacturers are required on most vehicles to be able to turn the MIL on if a leak occurs that is above .040". This is called *large leak detection*. Once the large leak detection has been accomplished, the system retests itself and looks for leaks that are around .020". These are really small leaks to find, and this is part of the reason why EVAP is the toughest monitor to run. In general terms, some systems must first allow engine vacuum to pull a vacuum on the entire fuel system, including the tank. Once the system is in vacuum, it is sealed and observed over a period of time. Leaks allow the vacuum to escape, and the faster it escapes, the larger the leak. Note that some vehicles test using pressure rather than vacuum. As the vehicle moves along the road, the fuel sloshes, changing the volume above the gas. Frequently this shuts the monitor off, possibly requiring an eight-hour soak before it will run again. The enabling criteria are the most complicated and difficult to attain for EVAP. The scan that shows the EVAP not completed is a very common situation (Figure 5–10).

OBD II Drive Cycles	Lexus
	All Models

EVAP (Vacuum Pressure, Intrusive Type)	2004–2005
	All Engines

Notes	1. Monitor status will be rest to "Incomplete" if the ECU loses power, DTCs are cleared or conditions for enabling conditions have not been met. 2. If a drive cycle is interrupted, it can be resumed. 3. Avoid sudden speed changes.
Conditions	1. MIL must be OFF; altitude <7800 feet. 2. Fuel level between 1/2–3/4. 3. IAT & ECT 40–95 degrees F; difference between IAT and ECT at start-up <13 degrees F.
Step 1	A cold soak must be performed prior to performing this drive cycle.
Step 2	Let the vehicle cold soak for 8 hours or until difference between IAT and ECT is less than 13 degrees F.
Step 3	This drive cycle can be performed at temperatures below 40 degrees F or at altitudes above 7800 feet, if the complete drive cycle, including Cold Soak, is repeated a 2nd time after cycling the ignition OFF.
Step 4	Connect scan tool and check monitor status and preconditions.
Step 5	Release the pressure in the fuel tank by removing and re-installing the fuel cap.
Step 6	Start the engine and let it idle until the ECT is at least 167 degrees F.
Step 7	Run the engine at 3000 rpm for approximately 10 seconds.
Step 8	Idle with the A/C ON to create a slight load, for 15–50 minutes.
Step 9	If vehicle is not equipped with A/C, set the parking brake and block the drive wheels with wheel chocks. Idle in Drive for 15–50 minutes.
Step 10	Check the monitor status. It should switch to "Complete". If not, turn the ignition OFF, be sure enabling criteria are met and repeat Steps 1–8.

Delmar/Cengage Learning

Figure 5–10 Frequently the most complicated enabling criteria are for the EVAP monitor.

Probably the most difficult situation that the technician faces is the DTC indicating an EVAP leak or flow problem. This indicates that apparently the monitor ran. The technician repairs the leak and is required to turn off the MIL or the rest of the monitors will not run. This may require an eight-hour soak and keeping the vehicle overnight to get the monitors to run. This is not the best situation for either the technician or the customer.

The next two monitors deal with the heated oxygen sensor. There probably is no more important sensor than the O2Ss. Their signals are used for fuel trim issues and to determine the success of other monitors. The O2S monitors are usually pretty straightforward and relatively easy to get to run (Figure 5–11).

The O2Ss will be tested in at least four different methods by the PCM when the monitors run. The first test usually is for the heater function. Small heaters are placed within the O2S to provide almost instant on operation. The vehicle needs to be in fuel control very quickly, and the heater makes this possible. Because the heater is resistive, it is easy for the PCM to see a voltage drop across it. If the heater were to open, there would not be any current flow and therefore no voltage drop. Once the heater is functional, a dynamic test is performed on the sensor. Dynamic means that the PCM forces the fuel system to run rich and then lean while it monitors the signal. When it forces the fuel injectors rich, the oxygen sensor's signal should rise and go sufficiently high to indicate the rich condition.

If the O2S passes the rich voltage test, the fuel injectors are turned off for a split second. This forces the signal lean and allows the PCM to test the lean voltage.

The last piece of the puzzle is usually the speed test. The response to a changing rich or lean condition must be both fast and accurate. As the vehicle drives and the customer moves the accelerator pedal, the fuel system must add or take away fuel based on the conditions present. The PCM momentarily turns on the fuel injectors to test the lean to rich time, and turns them off to look at the rich to lean time. All of these tests must be passed for the monitor to run and not generate a DTC. You need to understand that the most important sensors for OBD II operation are the oxygen sensors. If their operation is not up to the standards required, other monitors might be affected or not run. In Chapter 10, we will see how to duplicate this series of tests in the oxygen sensor monitor.

The last non-continuous monitor that we need to look at is the EGR monitor (Figure 5–12).

The exhaust gas recirculation monitor will check the overall operation of the system but also its ability to flow exhaust. Under certain operating conditions the PCM needs to add some exhaust to the intake manifold. This already burned mixture is inert and will not support additional burning. It basically takes up space in the intake. This tends to reduce the production of nitrous oxide (NOx). Many vehicles have eliminated the use of the EGR system because they are able to control NOx through aggressive fuel control. These vehicles will have N/A shown on the monitor status screen of the scanner. Vehicles that actually have an EGR system will have to monitor it for flow

Figure 5-11 This oxygen sensor will be tested by the O2S monitor.

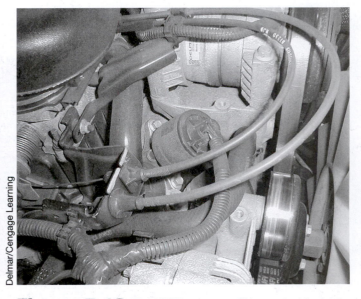

Figure 5-12 This EGR system will be tested for both function and flow.

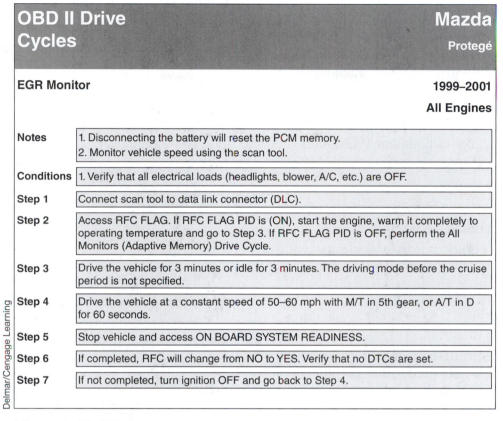

OBD II Drive Cycles		Mazda Protegé
EGR Monitor		**1999–2001**
		All Engines
Notes	1. Disconnecting the battery will reset the PCM memory. 2. Monitor vehicle speed using the scan tool.	
Conditions	1. Verify that all electrical loads (headlights, blower, A/C, etc.) are OFF.	
Step 1	Connect scan tool to data link connector (DLC).	
Step 2	Access RFC FLAG. If RFC FLAG PID is (ON), start the engine, warm it completely to operating temperature and go to Step 3. If RFC FLAG PID is OFF, perform the All Monitors (Adaptive Memory) Drive Cycle.	
Step 3	Drive the vehicle for 3 minutes or idle for 3 minutes. The driving mode before the cruise period is not specified.	
Step 4	Drive the vehicle at a constant speed of 50–60 mph with M/T in 5th gear, or A/T in D for 60 seconds.	
Step 5	Stop vehicle and access ON BOARD SYSTEM READINESS.	
Step 6	If completed, RFC will change from NO to YES. Verify that no DTCs are set.	
Step 7	If not completed, turn ignition OFF and go back to Step 4.	

Delmar/Cengage Learning

Figure 5-13 The EGR monitor is one of the easiest non-continuous monitors to run.

primarily. The vehicle manufacturer will add a sensor or look at the signal of another sensor and calculate EGR flow indirectly. Ford is the only manufacturer that actually looks at the amount of EGR through the use of a dedicated sensor, the DPFE (differential pressure feedback EGR). The signal off of the DPFE is in direct proportion to the amount of EGR. The rest of the manufacturers use the indirect method of calculating EGR. For example, General Motors generally looks at the signal off of the MAP (manifold absolute pressure) sensor to figure out EGR flow. As the amount of EGR is increased, the manifold vacuum drops slightly. The drop allows the PCM to calculate the quantity of EGR.

The EGR monitor is generally one that runs easily, but does require specific enabling criteria. Figure 5–13 shows the enabling criteria for EGR on a Mazda.

Notice that the conditions that allow it to run are really not all that difficult to attain. Normal driving should allow this monitor to run. The criteria that says use the scanner to verify monitor status (RFC flag) is very important and another reason to maintain control of the vehicle. Setting the monitors prior to giving the vehicle back to the customer is always a good idea. It prevents the customer from driving and suddenly, when the monitor runs, seeing the MIL on again. Even if it is for another DTC, the customer believes the problem is the same, because the check engine light is on again. If you have taken the time to run the monitors, then you have seen the MIL come back on and can revisit the system at fault. Whenever possible, maintain control of the vehicle.

CONCLUSION

In this chapter we looked at the various strategies required to get the monitors to run. We separated the monitors into continuous and non-continuous and examined the enabling criteria necessary to run to completion. We also showed how the freeze frame data can be useful in clearing the MIL and getting the vehicle through an emission test. We discussed how various States have reject criteria within their emission testing program. The use of fuel trim data to help diagnose a vehicle and/or verify a repair was examined in addition to using the enabling criteria to get a non-continuous monitor to run.

REVIEW QUESTIONS

1. Technician A states that the continuous monitors are the easiest to run since their enabling criteria are usually start and run. Technician B states that the non-continuous monitors will have enabling criteria that will involve specific driving sequences and inputs to run to completion. Who is correct?

   a. Technician A only

   b. Technician B only

   c. Both Technician A and B

   d. Neither Technician A nor B

2. Start and run criteria is usually associated with

   a. freeze frame

   b. non-continuous monitors

   c. EVAP monitor

   d. continuous monitors

3. Short-term fuel trim is being discussed. Technician A states that STFT is stored in the computer memory after the ignition key is turned off. Technician B states that STFT is a result of the oxygen sensor(s) signal to the PCM. Who is correct?

   a. Technician A only

   b. Technician B only

   c. Both Technician A and B

   d. Neither Technician A nor B

4. Driving the freeze frame is a method sometimes used to

   a. extinguish the MIL

   b. eliminate the DTC

   c. run the non-continuous monitors

   d. reset the fuel trim values to zero

5. Which monitor is most like OBD I?

   a. misfire

   b. comprehensive component

   c. EVAP

   d. fuel

6. Technician A states that the enabling criteria for the EVAP monitor is the easiest to run. Technician B states that the misfire monitor is the most difficult to run. Who is correct?

   a. Technician A only

   b. Technician B only

   c. Both Technician A and B

   d. Neither Technician A nor B

7. The enabling criteria call for a cold start or an eight-hour soak. This means that the

   a. ECT must be up to operating temperature at start-up

   b. IAT must be 50 degrees less than ECT

   c. keep-alive memory sees a minimum of eight hours since last start-up

   d. ECT and the IAT are close in temperature

8. Technician A states that States might reject a 1999 vehicle if it has three or more monitors not run. Technician B states that States might reject a 2002 vehicle if it has two or more monitors not run. Who is correct?

   a. Technician A only

   b. Technician B only

   c. Both Technician A and B

   d. Neither Technician A nor B

9. STFT is −18%. LTFT is +18%. This indicates that the vehicle

   a. is fixed

   b. will fail an emission test

   c. needs to have the monitors run to completion

   d. is running lean and trim in trying to correct for the condition

10. Non-continuous monitors are being discussed. Technician A states that a complicated set of enabling criteria might be required to run the monitors to completion. Technician B states that frequently start and run are the enabling criteria required to run the monitors to completion. Who is correct?

    a. Technician A only

    b. Technician B only

    c. Both Technician A and B

    d. Neither Technician A nor B

CHAPTER 6

SCAN DATA

OBJECTIVES

At the conclusion of this chapter you should be able to: ■ Scan for generic data ■ Scan for OEM data ■ Recognize the difference between generic and OEM data ■ Identify the various operational modes available within OBD II ■ Identify the DLC connector pin-out ■ Test the DLC for power and ground availability ■ Test the DLC for serial data

INTRODUCTION

In this chapter, we'll cover an important topic: scan data. Communication between the technician's scanner and the vehicle's PCM occurs easily if protocol is matched. We'll discuss how the communication protocol is the speed and type of transmission and varies from manufacturer to manufacturer. The protocol at the scanner and the PCM must match or no communication occurs.

COMMUNICATION ON THE GENERIC SIDE

When OBD II was introduced, generic communication speeds and protocol were locked into what is now very slow transmission of data. The OBD II system has the ability to deliver information along various data lines. Both the generic side of the PCM and the OEM side have the ability to deliver information. However, the speed and number of PIDs available are where the main difference occurs. Generic data is limited to a data speed established in 1996 and still continues today. However, the OEM side of the system has been allowed to increase both the number of PIDs and the speed of transmission. Let's look at generic first. Figure 6–1 shows a generic list of data for an early OBD II vehicle.

This is very typical of what you will see when scanning on the generic side of the system. There are some very important pieces of information listed here. The bottom of the left column shows the vehicle certification as OBD II. You might think that this is rather obvious, but remember not all vehicles produced worldwide in the early days of OBD II were

TECHVIEW

File Tester View mode Setup Window Help

Real-time Tester data

Engine Speed =	725 RPM	ECT =	87 °C
Vehicle Spd =	0 km/h	Ign. Timing =	14.5 °
Engine Load =	3.1 %	MAP (P) =	32 kPaA
MAF (R) =	5.6 gm/s	TPS (%) =	0.0 %
IAT =	24 °C	Fuel Stat 1 =	CL
Fuel Stat 2 =	CL	ST FT 1 =	-0.8 %
LT FT 1 =	4.7 %	ST FT 2 =	1.5 %
LT FT 2 =	2.3 %	O2S B1 S1 =	0.835 VDC
FT O2S B1 S1 =	0.0 %	O2S B1 S3 =	0.215 VDC
O2S B2 S1 =	0.925 VDC	FT O2S B2 S1 =	2.3 %
MIL Status =	Off	Stored DTCs =	0
OBD Cert. =	OBD II		

Delmar/Cengage Learning

Figure 6-1 Generic data with limited PIDs.

certified as OBD II. There were a tremendous number of Canadian and German vehicles produced and never should have made their way to the United States. These "gray market" vehicles would have great difficulty attaining US OBD II standards. Right above the certification PID is MIL Status, currently off. It is important to note that MIL status "lives" within the generic side of the PCM. This is the command status that we discussed regarding the check engine light and its relationship to emission testing failures. When a vehicle arrives for an emission test, the factor that determines pass or fail will be MIL command status. The test station will scan this PID to determine the status of the MIL. If it is on, the vehicle fails. If it is off, the vehicle is scanned for monitor status and a determination of pass or reject is made. When you begin working on a vehicle that has failed the emission test, always scan the generic side of the PCM. By doing this you are duplicating the test conditions. It is only the information found on the generic side of the system that is used to determine pass/fail status at the test lane. When you connect to the DLC, your scanner should give you choices like those listed in Figure 6-2. Note: the DLC is referred to by many definitions. Data line communication, data link connector, and even diagnostic link connector. It is the connector under the dashboard that allows a scanner to interface with the vehicle PCM.

This illustration shows just a portion of the choices available. The "global OBDII" line is the generic side of the system. The two choices within global OBD II, MT or T1, allow you to choose the type of display.

Figure 6-2 There will be choices for the selection of generic or OEM PIDs.

OBD II Operational Modes

- 1 = Datastream
- 2 = Freeze frame
- 3 = DTC's
- 4 = Clear DTC's and monitor status
- 5 = O2S data
- 6 = Non continuous monitor data
- 7 = Continuous monitor data
- 8 = EVAP
- 9 = Vehicle information

Figure 6-3 Operational modes within the global or generic side of OBD II.

MT is MasterTech mode, and T1 is Tech One mode. Either will deliver the generic information that you need. Within the generic side of the system will be the various modes available. Figure 6-3 shows the Operational modes.

The data we have been looking at is in operational mode #1, datastream. The other modes are self-explanatory. If a DTC is fired off, it will show in mode 3, with freeze frame data in #2. Notice modes 5, 6, and 7. Monitor functions are found here. These modes are very helpful when diagnosing, repairing, and running monitors on vehicles and will be covered in depth in future chapters.

If we go back to the data listed in Figure 6-1, you will notice that there are few PIDs available: only 23. Realistically, this isn't enough information to effectively repair many vehicles. There is sufficient information here, however, to run an emission test. The important thing about the generic side of the PCM is what data is available and stored here. The generic side has monitor status, DTC (at least the P0s) certification, MIL status and fuel trim values. You must have access to this side of the system if you will be doing emission repairs. Some scanners only show the generic side, while others only show the OEM side. What scanner you or your shop chooses is up to you. However, remember that you need access to both sides of the system, which may mean two separate scanners. Once you are "in" the generic side of the system, additional functions will be available. Figure 6-4 shows the submenus available.

You can see that there is a data list available (our limited one from Figure 6-1), DTCs and their snapshots can be read, and the all-important OBD evaluations scanned. OBD evaluations are where we would find the monitor status screen that we have looked at

Figure 6-4 Various functions available on the generic side of the PCM.

in a previous chapter. Everything the State requires to pass or fail a vehicle is available on the generic side of the PCM.

On our example of an early OBD II GM product we hade 23 PIDs. If we go back to the main menu (Figure 6-2) and instead of choosing the global (generic) side we choose the GM P/T, or General Motors powertrain, we would enter the OEM side of the system. Figures 6-5 and 6-6 show the PIDs available.

Notice that there are some substantial differences. The most obvious one is the fact that we now have 73 PIDs, which cover additional systems such as injector pulse width, system voltages, EGR function, MAF (mass air flow), EVAP functions, etc. Remember earlier that we mentioned that the MIL, certification, and monitors would be found only on the generic side? Notice that MIL command status and certification are not found on the OEM scan data list of PIDs. This is why it is so important to always look at the generic side of the system when faced with an emission failure. Failures are based off of generic data, but many diagnostic functions are on the OEM side.

The other thing to always keep in mind is data speed. Generic data is at 1996 speed and will not change. OEM data has kept up with the computer industry and gotten faster and faster. Fast data (OEM) can be used for graphic displays that some scanners support. It is possible to look at oxygen sensors or switching valves or other forms of relatively fast data straight off of the appropriate PID. The display will update fast enough for accurate diagnosis. Generic data is generally too slow for graphic functions.

DATA COMMUNICATION VIA THE DLC

The connection to the PCM is accomplished at the DLC and, like many other OBD II functions, its shape, location, and pin-out are standardized. Figure 6-7 shows the DLC.

This 16-pin connector contains both generic and OEM pins. Seven of the pins are standardized and are the same on all vehicles. Nine of the pins are the manufacturer's discretionary pins, which the OEM can use for any purpose. These nine pins deliver the

Available GM Scan Data 1

TECHVIEW

File Tester View mode Setup Window Help

Real-time Tester data

Engine Speed =	635 RPM		Spark Advance =	19 °
IAC =	62		TPS (V) =	0.54 V
TPS (%) =	0 %		Vehicle Speed =	0 MPH
ECT (°) =	86 °C		IAT (°) =	27 °C
Loop Status =	Closed		Engine Load =	3 %
MAP (P) =	32 KPa		MAF (R) =	4.7 g/s
BARO (P) =	99 KPa		Ignition (V) =	14.1 V
Desired IAC =	62		Desired Idle =	625 RPM
Rich/Lean Stat 1 =	Lean		Rich/Lean Stat 2 =	Lean
Long Term FT B1 =	134		Long Term FT B2 =	131
Short Term FT B1 =	127		Short Term FT B2 =	127
Air Fuel Ratio =	14.7		KS Active Cntr. =	133
EGR Duty Cycle =	0 %		EGR Position =	0 %
EGR Desired Pos =	0 %		Evap Purge D.C. =	17 %
Engine Run Time =	0:07:22		Startup ECT =	75 °C

Figure 6-5 OEM scan data available for an early GM.

Available GM Scan Data 2

TECHVIEW

File Tester View mode Setup Window Help

Real-time Tester data

Engine Speed =	634	RPM	IAC =	61	
TPS [V] =	0.54	V	TPS [%] =	0	%
Vehicle Speed =	0	MPH	ECT [°] =	87	°C
IAT [°] =	26	°C	Engine Load =	3	%
MAP [P] =	32	KPa	MAF [R] =	4.7	g/s
BARO [P] =	99	KPa	Ignition [V] =	14.1	V
Desired IAC =	60		Desired Idle =	625	RPM
Rich/Lean Stat 1 =	Lean		Rich/Lean Stat 2 =	Rich	
Injector P/W 1 =	3	mS	Injector P/W 2 =	3	mS
Air Fuel Ratio =	14.7		EGR Duty Cycle =	0	%
EGR Position =	0	%	EGR Desired Pos =	0	%
Evap Purge D.C. =	17	%	A/C Request =	No	
A/C Relay =	Off		Engine Run Time =	0:09:14	
Startup ECT =	75	°C	TCC Duty Cycle =	0.00	%
1/2 Shift Sol. =	On		2/3 Shift Sol. =	On	
Brake Switch =	Closed		TCC Enable =	Yes	
Evap. Vent Sol. =	Off				

Figure 6-6 The rest of the OEM data available for an early GM.

SAE J1962 Connector

1	2	3	4	5	6	7	8
9	10	11	12	13	14	15	16

Pin Assignments

PIN 1 - Manufacturers discretion
PIN 2 - SAE J1850 Line (Bus +)*
PIN 3 - Manufacturers discretion
PIN 4 - Chassis Ground
PIN 5 - Signal Ground
PIN 6 - SAE J2284 (CAN High)*
PIN 7 - K Line of ISO 9141-2 & ISO/DIS 4230-4*
PIN 8 - Manufacturers discretion
PIN 9 - Manufacturers discretion
PIN 10 - SAE J1850 Line (Bus –)*
PIN 11 - Manufacturers discretion
PIN 12 - Manufacturers discretion
PIN 13 - Manufacturers discretion
PIN 14 - SAE J2284 (CAN Low)*
PIN 15 - L Line of ISO 9141-2 & ISO/DIS 4230-4*
PIN 16 - Unswitched Vehicle Battery Positive

Figure 6-7 All OBD II connectors for scanners have the same pin-out.

OEM data. It is also important to note that pin 16 has an unswitched power, while pin four is a chassis ground. Many scanners will pick up power and ground from pins 16 and four. This allows them to receive communication without requiring a power cable, as was necessary on OBD I vehicles. Notice on pins 2, 6, 7, 10, 14, and 15 there are some strange letters and numbers. This is the communication protocol that the manufacturer chooses. Data can be transmitted in a variety of protocols, and your scanner will initially scan for the available protocols and lock on to the one chosen by the manufacturer. J-1850, J-2284, ISO-9141-2, and ISO/DIS 4230-4 are the common communication protocols available. Some scanners will indicate the communication protocol on the screen once they have locked into what the vehicle is using.

One additional point regarding the DLC. When an emission test is being done on a vehicle for State inspection purposes, pin 16 must have B+ (battery positive), and pin 5 (signal ground) must be B- (battery negative). We sometimes find this to be a problem when the test station indicates no communication and fails a vehicle. For example, the State included the following explanation on the Repair Diagnostic Report in Figure 6–8: "Failure reason: We were unable to communicate with your vehicle's OBD system. The OBD system might be inoperative." The Kia failed because of lack of communication at the emission test lane, so the customer took the vehicle to a shop.

The shop plugs in their scanner and the vehicle communicates. The battle between the State and the technician starts with the customer in the middle.

Delmar/Cengage Learning

OBD - REPAIR DIAGNOSTIC REPORT (RDR)

This Report contains information that will help a repair technician diagnose and repair your vehicle.

Plate: VIN: VIR: V016182357 Make: KIA Model: SPECTR Model Year: 2001	Test Time: 02/07/2006 16:36:25 Test No: 3 Vehicle Type: LDV, S Test Type: OBDII Station: 27 Lane: 1	MIL Commanded: Bulb Check: PASS Ready Status:

OBD TEST RESULT - FAIL

Your vehicle's computerized self-diagnostic system (OBD) registered the fault(s) listed below. This fault(s) is probably an indication of a malfunction of an emission component. However, multiple and/or seemingly unrelated faults may be an indication of an emission-related problem that occurred previously but upon further evaluation by the OBD system was determined to be only temporary. Therefore, proper diagnosis by a qualified technician is required to positively identify the source of any emission-related problem. The problems identified on this Repair Diagnostic Report must be corrected in order to pass the required OBDII test.

FAILURE REASON: WE WERE UNABLE TO COMMUNICATE WITH YOUR VEHICLE'S OBD SYSTEM. THE OBD SYSTEM MAY BE INOPERATIVE.

Delmar/Cengage Learning

Figure 6-8 A lack of communication caused this vehicle to fail.

The scanner that the tech plugged in requires pin 16 for power and pin four for ground. The State emission test scanner uses power from pin 16 *but* ground from pin five. This is where the problem occurs. When you are faced with a communication problem, always start with a test for power on pin 16 and grounds on pins four and five. The most frequent reason why communication is not available is in the power and ground sides of the DLC.

If you have power and ground available, the next most logical place to look will be in the datastream. For this, back probe the DLC communication pin and sensor ground (#5) with a DSO (digital storage oscilloscope) and turn the key on with the engine off. Communication will look like a series of very fast on/off signals, such as those shown in Figure 6–9.

This series of on and off signals will be extremely fast. Only a DSO will be capable of looking at it. The Kia in question had communication on the data line but was missing the pin #5 ground, resulting in the non-communication error failure. Replacing the pin and running a new ground cured the problem. This vehicle had failed the emission test and was unable to get a valid license for 8 months, and the only thing wrong was a broken ground. Don't forget, it takes power on pin 16, a ground on both pins four and five, plus available serial data for communication to occur. Also, remember that your scanner will use ground pin four while the emission test will use pin five. This

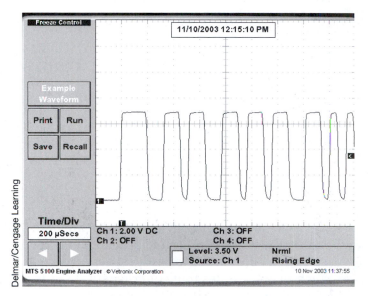

Figure 6-9 Serial data at the DLC will look like a series of very fast on/off signals.

ground difference is a frequent cause for discussion between a technician and the State EPA.

CONCLUSION

In this chapter we looked at the ability of the OBD II system to communicate through a scanner. We examined the differences between the generic side and the OEM side of the data and viewed what data is available. We also looked at the various pins within the connector and discussed how important the power and ground pins are for emission testing. We learned how to use the DSO to indicate communication availability in case of an emission failure due to lack of communication in the test lane.

REVIEW QUESTIONS

1. Technician A states that the generic side of scan data will be used by the State emission tests to determine pass/fail/reject status. Technician B states that the OEM side of scan data will not have as much data available as the generic side. Who is correct?

   a. Technician A only

   b. Technician B only

   c. Both Technician A and B

   d. Neither Technician A nor B

2. The certification status of the vehicle is being discussed. Technician A states that OBD certification (OBD I or OBD II) will be found on the OEM side. Technician B states that MIL status will be found on the generic side of the scan data. Who is correct?

   a. Technician A only

   b. Technician B only

   c. Both Technician A and B

   d. Neither Technician A nor B

3. PIDs are

   a. programmable identifiers

   b. positive indicated determiners

   c. positive internal directions

   d. parameter identifiers

4. Generic data is being discussed. Technician A states that it is rather limited and might not give sufficient information to easily repair the vehicle. Technician B states that generic data should be viewed prior to returning a failed vehicle to an emission test station. Who is correct?

   a. Technician A only

   b. Technician B only

   c. Both Technician A and B

   d. Neither Technician A nor B

5. OBD II operational modes are the

   a. sections or places where data is available

   b. modes that a driver uses when driving to get monitors set

   c. places where data is stored regarding the emission test

   d. none of the above

6. Which mode contains the live scan data?

   a. mode 9

   b. mode 2

   c. mode 3

   d. mode 1

7. The OBD II connector is

   a. different for each manufacturer

   b. the same for all manufacturers

   c. located under the hood of the vehicle

   d. where only a manufacturer-specific scanner can be connected

8. Technician A states that the manufacturer's discretionary pins within the connector are where the generic data will be available. Technician B states that pins six and 12 have power and ground available. Who is correct?

   a. Technician A only

   b. Technician B only

   c. Both Technician A and B

   d. Neither Technician A nor B

9. A State emission test indicates failure because of a lack of communication. This indicates that

   a. power might be missing from the connector

   b. ground might be missing from the connector

   c. data might be missing from the connector

   d. all of the above

10. A DSO shows a flat 0 V line when connected to the OBD II connector data pin and ground. This indicates

    a. nothing

    b. normal activity

    c. that scan data is available

    d. there is no communication available on the data pin

CONTINUOUS MONITORS—MISFIRE

OBJECTIVES

At the conclusion of this chapter you should be able to: ■ Identify the inputs that are used for misfire detection ■ Recognize the causes of some false misfire DTCs ■ Identify the pieces of the DTC used for misfire ■ Capture and analyze a compression waveform ■ Capture and analyze an ignition waveform ■ Capture and analyze a fuel injector waveform ■ Compute fuel pump speed off of a DSO waveform

INTRODUCTION

It is now time to begin the analysis of each monitor. We will look at the enabling criteria, the function, and how the monitor works in detail. The first continuous monitor that we will examine is the misfire monitor.

MISFIRE MONITOR

There are three continuous monitors that generate DTCs. The misfire monitor generates more DTCs than the other two, so it makes sense to start with it. The function is self-explanatory; this monitor will detect misfires and generate a DTC that identifies both the fact that a misfire is occurring and frequently the cylinder number. So a P0301 indicates that cylinder #1 has been misfiring, while a P0300 indicates a general random misfire involving more than one cylinder. Say, for instance, that an intake manifold leak causes three cylinders of a V-6 engine to misfire. When the vehicle is scanned, you might see a P0300, plus a P0301, 302, 303. It is not uncommon for the PCM to generate the random misfire when it sees misfire for half of the cylinders. Note that not all manufacturers support the individual cylinder misfire DTC. Most, however, do.

Let's look at the how of misfire detection. As an engine runs, it is constantly changing speeds—not only because of the driver moving the gas pedal, but because of the individual power impulses that occur

from each cylinder. Figure 7–1 shows a V-8 engine firing for four full rotations of the crankshaft.

Notice that the actual crankshaft speed varies between 1580 and 1600 RPM. This changing engine speed is because each cylinder, as it fires, will increase the speed, and in between the firings, the engine will slow down again. These very slight engine speed changes are not seen or felt by the driver and are virtually impossible to see with a digital storage scope, scanner, or tachometer. However, they are seen by the PCM, which monitors them continuously. There is not a set of enabling criteria required to get the misfire monitor to run. It is running all of the time. If a single cylinder were to misfire just once over the four crankshaft rotations that the illustration shows, would the PCM be capable of "seeing" it? The answer is yes.

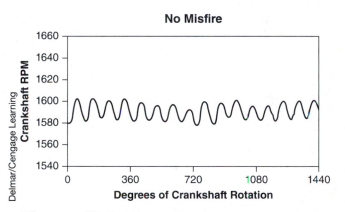

Figure 7-1 A V-8 crankshaft position sensor's signal.

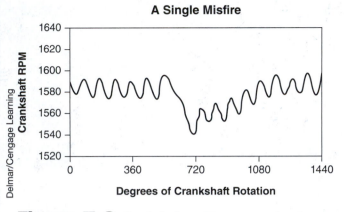

Figure 7-2 Crankshaft position sensor signal showing misfire.

Figure 7-2 shows the same PCM RPM trace with a single cylinder misfiring just once.

Engine speed has dropped from a high of 1600 to a low of 1540 and then, as additional cylinders fired, the speed has increased to the point it was prior to the misfire. This speed trace shows four full rotations of the crankshaft. Each 360 degree (or one) rotation has four power pulses, and only once did one cylinder misfire. This would not generate a DTC, because it is only a one-time event; however, if the pattern continued, eventually the PCM would generate a DTC that would steer you to the correct cylinder. A misfire DTC like a P0302 will show that cylinder #2 is misfiring, but will not indicate why the misfire is occurring. We will see later in this chapter than this means we need to go back to the basics of compression, ignition, and fuel to determine the cause. Additionally, there are some scanner modes that might indicate that a misfire has occurred over a series of crankshaft rotations. Figure 7-3 shows a scan from a GM vehicle without misfire.

Remember that the scanner will lead you to the correct cylinder, but might not give you all the information that might be required to actually repair the vehicle.

It is very important to note that the PCM looks at the spacing of the pulses of the CKP to determine misfire. This will only be done under conditions of the throttle not changing. Basically the engine must not have a variable RPM. Typical systems look at the TP (throttle position sensor) for a constant or steady voltage and then look at the CKP. If the CKP shows changing speeds with a steady throttle, misfire has been detected. There is always a possibility that the PCM will see a misfire when actually there is none. Vibrations are common causes of misfire DTCs, where there is no "real" misfire.

It is also important to understand how the PCM knows that misfire has occurred. We previously mentioned that the PCM monitors engine RPM, but just how does it do that? On all engines there is an engine speed sensors usually referred to as the CKP or crankshaft position sensor. The job of the CKP is to deliver engine speed to the PCM. It is the CKP that will indirectly determine if a cylinder is misfiring and which cylinder it is. As an example, let's use a Ford CKP, which is located on the front of the engine. Figure 7-4 shows this sensor.

The vibration dampener has the trigger wheel, and the trigger wheel has a series of slots that, with one exception, are spaced 10 degrees apart from one another. As the engine turns, the variable reluctance sensor generates an AC signal. Figure 7-5 shows this signal.

Notice the missing-tooth region: this different amplitude of the signal results from the trigger wheel having 35 teeth spaced 10 degrees apart with one additional space. The space allows the CKP to generate a signal for a longer period of time, resulting in the spacing and amplitude difference. When the PCM

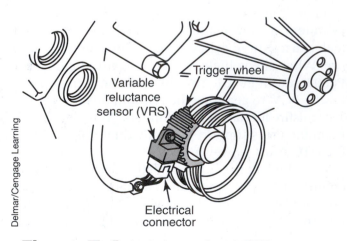

Figure 7-3 Scanner showing misfire data.

Figure 7-4 A missing-tooth-style CKP.

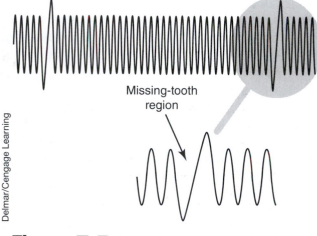

Figure 7-5 A generated AC signal that identifies piston position.

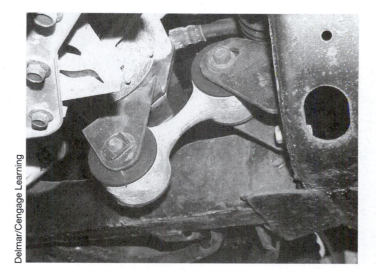

Figure 7-7 Torque struts are designed to prevent vibrations.

sees the missing-tooth signal, it knows the position of the crankshaft. When this signal is analyzed with the additional input from the cam sensor, the PCM can calculate which cylinder is misfiring.

The process of identifying a misfire and noting which cylinder is at fault is difficult. Sometimes the PCM determines a misfire when one does not exist. Anything that is perceived as the crankshaft speeding up or slowing down may be "tagged" as a misfire. There are numerous examples of incorrect P0300s being generated, especially in early OBD II vehicles. One of the most common causes of these false DTC are spinning components, such as water pumps, idler pulleys, power steering pumps, AC compressors, etc. The belts can also cause false DTCs (Figure 7–6). If the serpentine belt is missing some of the ribs that help guide it, the belt causes a slight speeding up and slowing down that is incorrectly labeled a misfire. Always check the belt and spinning components.

Another cause of false P0300 DTCs are vibrations. The vibrations cause the CKP to "see" a misfire when it might not exist. Figure 7–7 shows a torque strut that has the job of preventing vibration. If it is excessively worn, it cannot do its job. There are many examples of the end bushings getting so worn that the strut has metal-on-metal contact. Add this vibration and the CKP might see it as a misfire and set a P0300, random misfire DTC. Note that most false misfire DTCs will be of the random or P0300 variety. Additionally, if the PCM delivers misfire data on a cylinder-by-cylinder basis, you will see that no one cylinder will have a larger number of misfires than any other cylinder.

Another common cause of false misfire DTCs are the engine, transmission, or cradle mounts. Figures 7–8 and 7–9 show common setups that need inspection on a regular basis.

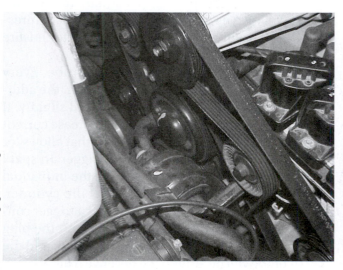

Figure 7-6 Belts can cause false P0300s.

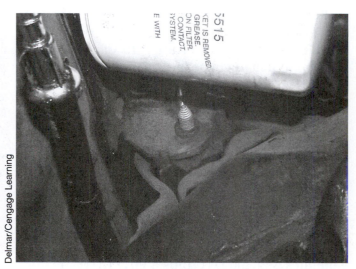

Figure 7-8 Engine mounts should be inspected.

Delmar/Cengage Learning

Figure 7-9 Cradle mounts help prevent vibrations.

Delmar/Cengage Learning

Figure 7-10 Any spinning component could cause a false P0300.

Remember, anything that is spinning and causing either an imbalance or a vibration could be the cause of a false misfire. Look at Figure 7–10. Can you identify a few items that might be the cause of a false DTC?

Between the belt, the pulleys, a water pump, an alternator, an idler pulley, and the supercharger, this vehicle has many items that might make false misfire DTCs relatively common but possibly difficult to identify.

ADDITIONAL CLUES TO MISFIRE

There are examples of misfire detection correctly identifying a misfire where we see additional clues or DTCs that help guide us. For instance, a vehicle generates a P0300 DTC plus a P0174 DTC. Is there additional information that could be helpful? Always examine all DTCs and look for cause and effect relationships that might steer you to the diagnosis and repair. A P0174 is an indication of a lean condition. So the obvious question is, can there be a relationship between the 174 and the 300? The answer is yes. If an engine is running very lean, we can pick up a lean misfire. The additional DTC has helped us to zero in on possible problems. In a later chapter, we will examine the use of fuel trim as a diagnostic aide. Our example of a 174 and 300 might be verified if the PCM were trying to correct the problem by adding fuel trim. Always scan for all DTCs.

FURTHER DIAGNOSIS

An engine is a pretty basic device. Take fuel and air, compress it, and ignite it with a spark and you have the basics of the internal compression engine. When an engine misfires, it is a problem with compression, ignition, or fuel. With the PCM in control of both fuel and ignition, it is sometimes difficult to identify which of the big three is at fault. Let's start with compression.

Compression testing can be done either mechanically with a compression gauge or indirectly with a current probe. To be practical, mechanical compression testing is a thing of the past. Many engines have made spark plug removal very difficult, and removal of the plugs is required for a mechanical compression test. Additionally, with the majority of engines being aluminum, spark plug removal becomes a problem. Most engine manufacturers want us *not* to remove spark plugs when the engine is hot, and compression testing is most accurate with a hot engine. Accomplishing the task becomes almost impossible. Electronic compression testing is the answer. Although it will not reveal PSI (pounds per square inch) numbers, it will give us a relative indication of compression. Relative will be sufficient, especially for misfire diagnosis.

We will use a DSO with a current probe. Every time a cylinder comes up on compression, the additional pressure slows down engine speed slightly. If you slow down a starter motor, it will increase current draw. It is this increasing current draw that allows for a relative compression test. Using a trigger on spark plug wire number one will also allow the individual current peaks to be related to a specific cylinder. Figure 7–11 shows a common DSO with a trigger connection. This trigger will be placed around the plug wire, the fuel injectors will be disconnected, and the current probe will be around the negative battery cable. Without fuel the engine will crank but will not start.

Figure 7-11 A DSO can be used to give relative compression results.

Let's start with a good engine. Figure 7–12 shows the results of a cranking relative compression test.

Notice that the peaks of the trace are all just about even. Indirectly the peaks say that the cranking compression is "relatively" the same. The "T" that shows on the bottom middle of the screen indicates where the trigger cylinder is located. Because the ignition pulse follows compression, the peak in front of the T is for cylinder #1, and then the peaks follow in firing order. The firing order for this engine is 1, 6, 5, 4, 3, 2. Cylinder #1 is left of center, and then 6, 5, and 4 follow (to the right of middle). Cylinders 3 and 2 are at the beginning of the trace on the left.

When we are analyzing and looking for the reasons why we have a misfire DTC present, a compression test makes a lot of sense. What does this compression

trace indicate? Simple: our misfire DTC is not being caused by compression problems. All of the peaks are equal, and as a result compression is also equal.

Let's take another example of a compression test. Figure 7–13 is from another V-6 engine. The DSO, current probe, and trigger have been set up the same and the engine is cranked. Do you notice any differences?

A peak is missing to the right of center. If the engine has the same 1, 6, 5, 4, 3, 2 firing order, what cylinder shows a different amount of current? Cylinder #5. Apparently, cranking current is lower for #5 than the rest of the cylinders, and this indicates a compression loss. It is important to note that this test will not tell you what is wrong with the cylinder. It just sends up the red flag indicating that there is a problem. The vehicle in Figure 7–14 generated a P0305. When the shop got the vehicle, the consumer had spent many dollars putting in new plugs, wires, and cleaning fuel injectors. It is impossible to say that the vehicle did not need these items, but they did not fix the problem. The check engine light remained on, the vehicle failed the emission test, and it was denied registration. The final repair involved doing a valve job, which cured the compression problem. The light turned off and the vehicle was emission tested, where it passed. With the passed emission test in hand, the consumer was able to get his vehicle registered. This is another example showing that doing a correct diagnosis for misfire DTC should involve compression, ignition, and fuel.

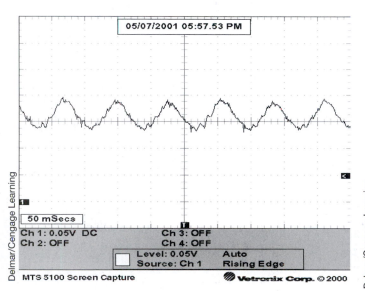

Figure 7-12 A DSO showing "good" relative compression.

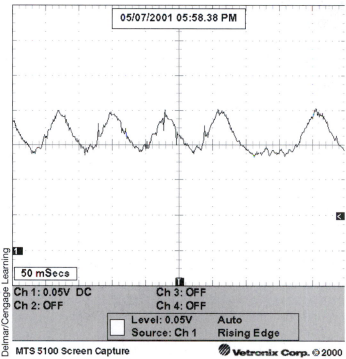
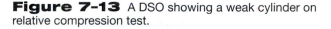

Figure 7-13 A DSO showing a weak cylinder on relative compression test.

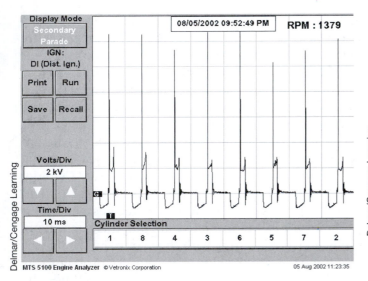

Figure 7-14 A DSO showing ignition firing voltage.

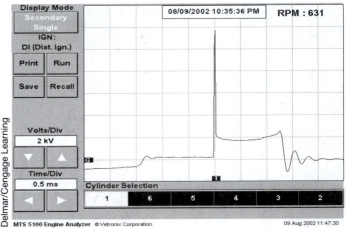

Figure 7-15 A single cylinder ignition trace.

IGNITION

This is not a text on ignition systems, so we will not look in great depth at the various systems that are on the road. They do share some common components, functions, and controls, however, that merit discussion. A simple DSO test of the ignition will reveal problems that might be the cause of misfire DTC. It really comes down to what was the firing voltage, how long did the spark last, and how long did primary current remain on? If these three functions are present at the correct level, then the ignition system is not the reason for the misfire DTC. Let's look at each one separately using our DSO.

Figure 7-14 shows the firing voltage for an 8-cylinder engine.

Each division is 2,000 volts (2 kV). The firing voltage on the engine runs from a low of 10 kV to a high of 13 kV. Most modern engines will fire at less than 15 kV and should have less than 3–4 kV difference between firings. We are at 10–13 kV, so this looks good. Remember that you should perform this test under the same conditions that freeze frame showed were present when the DTC was set. Try different engine speeds and loads.

The next important thing to look at is how long did the spark last? Figure 7-15 shows a single cylinder trace. The firing voltage is around 12 kV. The horizontal line immediately following the firing voltage spike in the middle of the screen indicates how long the spark actually lasted. This is called *burn time* or *burn duration,* and the general specification is to be greater than .8 mS. On this trace our time/div number is 0.5 mS, and the spark line extends for 2½ divisions, so this spark lasted for about 1.25 mS, well above the .8 mS minimum. It is during this time that the ignition system will get the mixture in the cylinder to burn.

Remember that this trace represents only one cylinder. We will need to look at each cylinder. Any cylinder that has a burn time less than .8 mS could cause a misfire DTC. Most manufacturers use multiple coils, with many using one coil per cylinder, so make sure you test each coil separately.

The last thing to look at is how long did the ignition system leave primary current on to the ignition coil? This charging of the coil is what allows the coil to ignite the mixture.

Figure 7-16 shows an ignition system, beginning with primary current on signal on the left side of the trace. The downward turn of the trace indicates when primary current begins to flow through the coil. Depending on the ignition system, primary current will flow up until the point of the spark or be cut back (current limit) once the coil is charged sufficiently. Our example has current limit. It shows as an upward

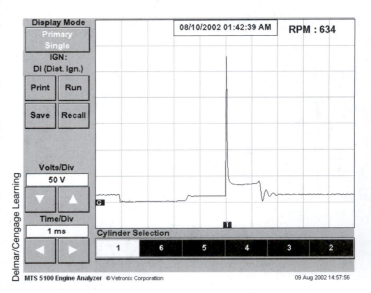

Figure 7-16 A primary pattern showing 2.5 mS primary "on" time.

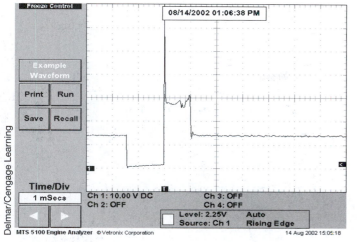

Figure 7-17 A primary pattern showing 1.5 mS primary "on" time.

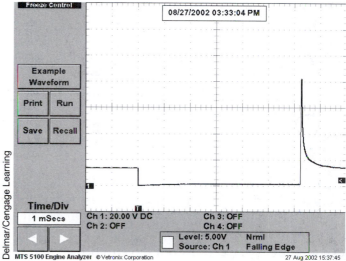

Figure 7-18 One-cylinder fuel injector pattern.

turn of the trace in the middle of the fourth (from the left) division. This primary on time, from primary turn on until current limit, should exceed 2.5 mS. Our example has adequate primary on time.

Figure 7–17 is from a vehicle with a P0300 DTC.

Do you notice anything suspicious? The DSO is again set for 1 mS per division. Look at the primary on time. It begins 1½ divisions from the left and runs until the firing line at the beginning of the third division. This primary on time is only 1.5 mS. With the minimum being 2.5 mS, this trace shows why the misfire occurred. This reduced primary on time also is likely the cause of the reduced burn time. Even though the minimum is 0.8 mS and our total is around 1.0 mS, notice that on the spark line the voltage begins to break up around 0.5 mS. It is likely that the reduced primary on time is the cause of the reduced burn time as these two functions go together.

FUEL INJECTION

The subject of fuel injection is, like ignition, beyond the scope of this text. We should recognize, however, that like ignition, there are some basic things that should be present. These items can be looked at using the DSO again. Figure 7–18 shows the voltage fuel injection pattern for one cylinder.

The voltage pattern shows the fuel injector being turned on at the beginning of the second division. The injector remains on, or flowing fuel, for over six divisions before it turns off. The spike toward the right side of the screen indicates the point where the PCM turned off injection. This is the *induced voltage* or *inductive kick* as it is frequently called. Remember that the DSO pattern indicated what happened electrically. We still have fuel pump pressure and volume

to consider, and both are very important to the function of the OBD II system. Without sufficient pressure the volume of fuel that is injected might be incorrect. Having the incorrect amount of fuel injected can cause problems that are not impossible for the system to overcome, but it will involve a fuel correction through the use of a function called *fuel trim,* which we will look at in detail in Chapter 8. It is always better for the system if it can calculate the correct amount of fuel and then inject it, rather than inject the wrong amount and have to adjust it with fuel trim.

Fuel pressure is easily measured with a fuel pressure gauge like the one shown in Figure 7–19.

This is usually a very important early step to take during any diagnosis. If the pressure is not correct, either the fuel pressure regulator is not correct and needs to be replaced or the fuel pump is unable to supply the minimum amount of fuel (volume) required. Insufficient fuel volume will result in lower

Figure 7-19 Fuel pressure gauge.

than normal pressure under load conditions. You can see that fuel pressure and volume are tied together. Without volume there cannot be adequate pressure, and without adequate pressure there cannot be good injection.

There are numerous tools available that will measure both pressure and volume. Figure 7–20 shows a common one. It is installed by opening up the system and connected in series so that all fuel flows through the tester.

Another method available is to use a DSO with a current probe connected into the electrical circuit. The analysis using a DSO is a bit more complicated than the direct method but has the advantage that the fuel system does not have to be opened to use. The current probe, such as the one shown in Figure 7–21, is placed around the electrical wire either supplying current to the fuel pump or supplying the ground.

The probe must be closed completely around the wire, and there can only be one wire in the probe. The fuel pump must be running and the DSO set for measuring AC voltage. As the pump spins it will both draw current and generate AC voltage. The AC voltage will be shown on the screen as a distinctive

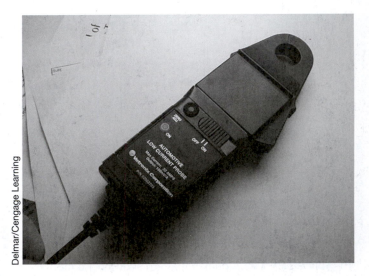

Figure 7-21 A current probe can be used to measure fuel pump speed.

pattern that can be analyzed for speed. There is a direct relationship between the speed of a pump and its volume. Figure 7–22 shows the pattern off of a fuel pump with the vehicle running. Follow along as we calculate the speed of this pump.

Each division of time, from left to right equals 2 mS or .002.

A millisecond is a thousandth of a second. Each downward turn of the trace is a commutator bar generating a voltage pulse. Each commutator bar will be electrically different. The pattern indicates which

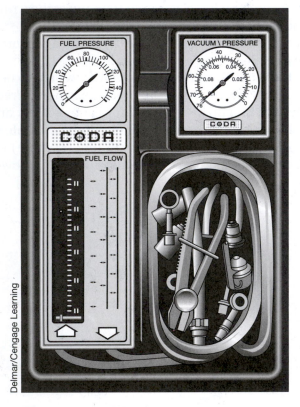

Figure 7-20 A combination fuel pressure and volume tester.

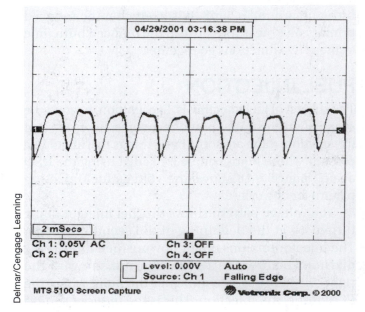

Figure 7-22 Each commutator bar has a distinct signal allowing fuel pump RPM to be calculated.

bar is in contact with the brush assembly of the fuel pump. Our example has eight commutator bars, which is very common. You can see that the first pattern and the ninth pattern are the same. A single rotation of the pump will generate the nine signals. The ninth is actually the first one again, and when we see it the pump has made one rotation.

The amount of time that this single rotation takes is our next concern. Remember each division or box indicates .002 sec. It looks like there are eight divisions, and eight times .002 equals .016 sec total. So it takes approximately .016 seconds for the pump to make one rotation. The next thing we need to calculate is how many times the pump rotates during a second. This is done by dividing one (representing a second) by the .016 sec. This pump rotates 62.5 times every second, and if we multiply this by 60 sec, we will have the speed or RPM of the pump. 62.5 × 60 sec = 3750 RPM. If you got a bit confused with the math, feel free to review it until it makes sense. Use a calculator to insure accuracy.

The analysis of this pump speed involves a simple number. The vast majority of the pumps in vehicles today need to spin a minimum of 4500 RPM to be able to produce sufficient volume for the fuel injection system to function correctly. Our example vehicle had a MIL on and two DTCs in memory: a P0300 (random misfire) and a P0174 (lean condition). Our analysis showed that the fuel pump was only spinning at 3750 RPM with a 4500 RPM minimum. It looks like this vehicle is running lean because the fuel pump cannot keep up with the volume required, and under certain load conditions the engine misfires. One last check for power and ground to the pump will show us if we need a new pump. If we are delivering good voltage under load (pump spinning) to the pump and the ground is also good, this pump needs to be replaced.

Don't forget that the OBD II system is designed to indicate via the scan data that a misfire is occurring. The P0300 indicates random misfire, while a P0305 indicates a misfire on cylinder #5. States that do emission testing find that misfire is extremely common and is usually in the top 20 DTCs. Additionally, some of the manufacturers will flash the check engine light when severe misfire occurs. This is supposed to alert the consumer that something is wrong and hopefully inspire him to bring the vehicle in for diagnosis and repair. Misfire that is allowed to occur unchecked will eventually destroy the catalytic converter. The misfire may allow unburned fuel to enter the CAT, where it will burn. This additional burning may increase the temperature of the CAT beyond its design limits and lead to its failure.

CONCLUSION

Misfire DTCs are among the most common DTCs. Vehicles that are emission tested will frequently find that within the top 20 DTCs are individual cylinders one through six plus the random misfire (P0300).

REVIEW QUESTIONS

1. Technician A states that misfire DTCs are generally set using the signal off of the CKP Technician B states false misfire DTCs may result from vibrations. Who is correct?
   a. Technician A only
   b. Technician B only
   c. Both Technician A and B
   d. Neither Technician A nor B

2. A CKP shows a drop in crankshaft speed that is always in the same place. This indicates
   a. a random miss is occurring
   b. a specific cylinder is misfiring
   c. the engine is running lean
   d. fuel pump speed is reduced

3. A P0305 DTC is set and observed with a scanner. Technician A states that this indicates a misfire problem with cylinder #4. Technician B states that this indicates a random misfire. Who is correct?
   a. Technician A only
   b. Technician B only
   c. Both Technician A and B
   d. Neither Technician A nor B

4. Technician A states that to capture a compression waveform one needs to install the current probe around the negative battery cable. Technician B states that the engine should be cranked over for a minimum of 15 sec to capture the compression pattern. Who is correct?
   a. Technician A only
   b. Technician B only
   c. Both Technician A and B
   d. Neither Technician A nor B

5. A scanner displays DTCs P0300 and P0304. Technician A states that this indicates a random misfire. Technician B states that this indicates a misfire in cylinder #4. Who is correct?

   a. Technician A only

   b. Technician B only

   c. Both Technician A and B

   d. Neither Technician A nor B

6. A compression test is being done on a V-6 engine. Which of the following is *not* required?

   a. shut down fuel delivery

   b. disable the ignition system

   c. connect a high current amp clamp around the negative battery cable

   d. crank the engine over for 15 seconds

7. An ignition waveform is being analyzed for an engine that displays a P0300 DTC. Technician A states that burn time should be less than .8 mS. Technician B states that primary on time should exceed 2.5 mS. Who is correct?

   a. Technician A only

   b. Technician B only

   c. Both Technician A and B

   d. Neither Technician A nor B

8. An injector waveform is being analyzed. All of the following are true statements except

   a. a current waveform can show when the pintle opens

   b. a voltage waveform can show when the pintle closes

   c. a voltage waveform can show the amount of time that the injector is open

   d. a current waveform can indicate the quality of the ground circuit

9. Technician A states that misfire DTCs can be caused by vibrations unrelated to actual misfires. Technician B states that the misfire monitor runs continuously. Who is correct?

   a. Technician A only

   b. Technician B only

   c. Both Technician A and B

   d. Neither Technician A nor B

10. A current waveform from a fuel pump is being analyzed. Technician A states that the waveform can be an indication of fuel volume. Technician B states that fuel pump RPM can be calculated. Who is correct?

    a. Technician A only

    b. Technician B only

    c. Both Technician A and B

    d. Neither Technician A nor B

THE FUEL SYSTEM MONITOR

INTRODUCTION

In this chapter, we will look at another continuous monitor, the fuel system monitor. You will see how the monitor functions within its enabling criteria, what it uses to accomplish the monitoring, and how the information can be used effectively.

This second continuous monitor is a very important one because the control of fuel delivery falls within the parameters. To understand the monitor we need to first look at both closed loop operation and the function of fuel trim. Let's start with closed loop operation. The function of the loop system is to take the information from the oxygen sensor(s) and use it to adjust the fuel delivery. If the O2S says the burn is

a bit lean, the PCM should add fuel. If the O2S says the burn is rich, the PCM should take away fuel. The reason this is called closed loop is because the burn produces an O2S signal, which is sent to the PCM and used to adjust the amount of fuel delivered; and once it is burned it will again produce an O2S signal. On and on the burn produces the O2S signal, Closed loop is an indication that the signal is being used, while open loop indicates that it isn't. With oxygen sensor heaters being common, the OBD II style of oxygen sensors begins to generate useful signals in very short order; in a matter of seconds in most cases. Figure 8–1 shows a common V-type engine with two sensors 1s. Each sensor 1 is being used to actually determine fuel

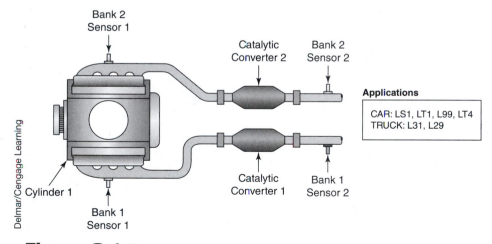

Bank 2
Sensor 1

Catalytic
Converter 2

Bank 2
Sensor 2

Applications

CAR: LS1, LT1, L99, LT4
TRUCK: L31, L29

Delmar/Cengage Learning

Cylinder 1

Bank 1
Sensor 1

Catalytic
Converter 1

Bank 1
Sensor 2

Figure 8–1 The position of the O2S determines its identification.

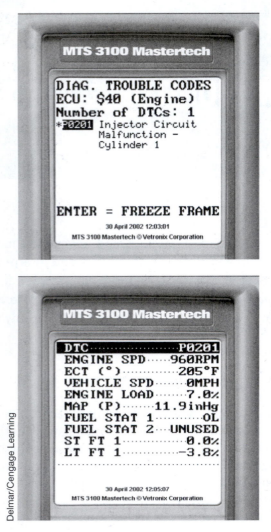

Figure 8-2 Freeze frame for a P0201.

delivery. Don't forget that bank 1 is the cylinder bank that has cylinder #1 in it, while bank 2 does not have cylinder #1 in it. The sensors labeled "sensor 2" are used to determine catalytic convertor efficiency. We will look at them in Chapter 11 on CAT.

When you are scanning a vehicle, always note loop status. It is basic to the operation of the system. Open loop, or "OL" on your scanner, indicates that something is wrong either with the sensor(s), the circuitry, or the PCM's use of the signal. It may also be present because the PCM has set certain DTCs. A misfire DTC on some vehicles will cause the PCM to put the system into open loop.

The other important thing to note when the vehicle is in open loop is that the data for fuel trim will not change or be inaccurate.

INJECTOR PULSE WIDTH

Injector pulse width is the main reason for the fuel monitor. The PCM will look at inputs regarding speed, temperature, and load at various speeds and convert

this information to a pulse width for the fuel injectors. If the conditions are light and warm, the pulse width will be low, as Figure 8-3 shows.

The downward turn of the trace that lines up with the second division is the beginning of the fuel injector pulse. The pulse ends with the upward turn of the trace just before the fifth division. According to the Time/Div setting (lower left), this DSO is set for 1 mSec (mS) per division, so this injector has remained on and flowing fuel for about 2.75 mSec (mS). This engine is fully warmed up, on a warm day, at idle speeds. If we increase the load on the engine by applying the brakes and increasing the throttle opening, we should see a corresponding increase in the amount of fuel injected. In other words, the pulse width should increase. Figure 8-4 shows the same vehicle under increased load.

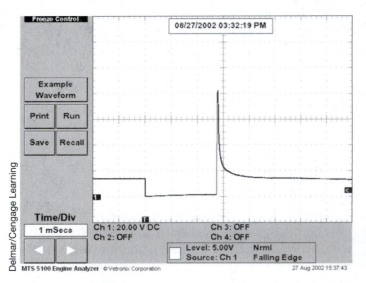

Figure 8-3 A fuel injector waveform with the engine at idle.

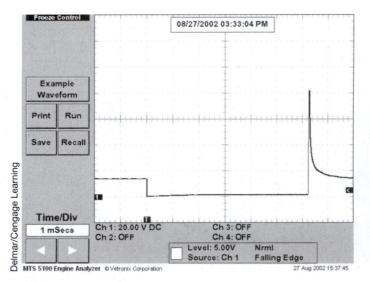

Figure 8-4 A fuel injector waveform with the engine under increased load.

Notice that the pulse width has increased to 6.25 mS. This increase is a function of the PCM and the fuel trim. This PCM will look at various inputs and determine the pulse width required. The oxygen sensor (O2S) will measure the results of fuel delivery and send a signal to the PCM of rich or lean. Based on the O2S's input, the PCM will adjust fuel delivery. This "fine tuning" of the fuel delivery is done by fuel trim.

FUEL TRIM

Fuel trim is a very important part of OBD II as it allows the PCM to make changes to the quantity of fuel injected based on the signal from the O2Ss. Usually only the front O2Ss, B1S1 and B2S1, are used if the vehicle engine is a V type and has two banks. The B1, which is part of B1S1, stands for bank 1, or the bank that has cylinder #1 in it. The B2 in B2S1 is for bank 1, or the bank that does not have cylinder #1 in it. The S1s signify the front or #1 oxygen sensors. The S1s are the sensors that are used to basically run the engine and help to determine fuel trim function. Simplified, the PCM determines the amount of fuel to inject based on various sensors. After the injected quantity of fuel burns, the information from the oxygen sensors is used to add or subtract fuel from the calculated injection amount. Adding or subtracting fuel is another way of saying to "trim" the quantity more toward the actual needs of the engine. Whether the engine is using oxygen sensors or air/fuel ratio sensors, the function of trim is the same: to adjust the quantity of fuel to the engine demands. There are two types of trim in use: short-term and long-term fuel trim.

Short-Term Fuel Trim   Short-term fuel trim, or STFT on your scanner, is an indication of how much fuel trim is needed to get the fuel system back into control. It is very fast and reactive, so it can make adjustments very quickly. It is the first trim to make adjustments when the quantity of fuel is inaccurate. For example, if a fuel injector were to get partially plugged and the amount of fuel injected to the one cylinder were reduced, STFT would kick in and add additional fuel to bring the vehicle back into fuel control. The key to understanding STFT is to recognize that it is used to make adjustments as soon as the O2S sees a problem. If the O2S voltage drops down, indicating a lean condition, fuel trim will increase the pulse width of the injector to compensate. If the O2S voltage rises up indicating a rich condition the STFT will adjust the pulse width down, leaning out the mixture until the O2S signal is normal. The adjustment to the fuel delivery is done and displayed on your scanner usually as a percentage either plus (+) or minus (-). Some manufacturers do not use the

percent of adjustment. Short-term fuel trim always precedes long-term fuel trim.

Long-Term Fuel Trim   LTFT is used in conjunction with STFT and, as its name implies, is over a longer period of time. It is a result of a problem that is fixed by STFT and needs to have a more lasting correction. It functions basically the same with the exception that it is based off of STFT. If STFT needs +8% to get into fuel control and the need is required over a long period of time, the STFT will be replaced by +8% LTFT. Let's look at an example of a 2.0L 4-cylinder OBD II engine in Figure 8–5.

Follow along and you will realize how STFT and LTFT actually function. With the vehicle running at a fixed 2000 RPM, we install a DSO on the B1S1 O2S and a scanner set to display STFT and LTFT PIDs. A PID is a single line of data transmitted from the PCM to the scanner. With the engine running, the O2S voltage is averaging 450 mV (.450V), which is considered perfect for this engine. At the same time, the PIDs for STFT and LTFT are fluctuating around 0%. Everything looks good. We now add propane (Figure 8–6).

Propane vs Trim

O2S	STFT	LTFT
450 mV	+/−0%	+/−0%

Delmar/Cengage Learning

Figure 8–5 An engine in fuel control.

Delmar/Cengage Learning

Figure 8–6 To artificially enrich the mixture, propane is added to the manifold.

The addition of the propane forces the system into a rich condition, resulting in the O2S voltage climbing up to greater than 800 mV (Figure 8–7).

Immediately, the STFT begins to compensate for the rich condition by taking away fuel ("going down"). LTFT is still around 0%. This brings up a question of why will STFT change and LTFT not? The answer is simple. LTFT has been programmed to not move until STFT is stable. With STFT moving down, LTFT will wait until the adjustment is completed. STFT is stored in a volatile memory, one that is powered by a key on battery source, so it will be lost should the engine be turned off. A key on is required for it to be functional. It is not stored or saved for the future.

Figure 8–8 illustrates our continuing trim analysis.

The third line indicates that the STFT has reached a stable value of −20%. It is compensating for the additional fuel being added—the propane—by taking away 20% of the pulse width. Notice that the vehicle is back in fuel control with a 450mV signal off of the O2S. Once everything is back in fuel control,

Propane vs Trim

O2S	STFT	LTFT
450 mV	+/−0%	+/−0%
550 mV	Going down	+/−0%

Delmar/Cengage Learning

Figure 8-7 The addition of propane drives the system rich.

Propane vs Trim

O2S	STFT	LTFT
450 mV	+/−0%	+/−0%
550 mV	Going down	+/−0%
450 mV	−20%	+/−0%

Delmar/Cengage Learning

Figure 8-8 Short-term fuel trim adjusts for the rich condition by taking away 20% fuel.

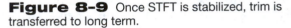
Propane vs Trim

O2S	STFT	LTFT
450 mV	+/−0%	+/−0%
550 mV	Going down	+/−0%
450 mV	−20%	+/−0%
450 mV	−10%	−10%
450 mV	+/−0%	−20%

Delmar/Cengage Learning

Figure 8-9 Once STFT is stabilized, trim is transferred to long term.

the PCM begins moving the fuel trim from STFT over to LTFT.

The fourth and fifth lines in Figure 8–9 show this transfer, so eventually all −20% is in LTFT. The sequence is important here. We started with a problem of additional fuel, which the PCM compensated for through STFT. Once sufficient STFT had the problem under control, the PCM transferred the trim over to the LTFT, where it would be stored even with key off. LTFT is powered by a keep-alive circuit that is hot or powered up all of the time, even in key off situations. Figure 8–10 shows the keep-alive circuit for a Ford vehicle.

Pin 55 circuit #1076 is labeled as Keep Alive Power (B+). This circuit has power all of the time, even with the key off. The function of this circuit is to maintain certain memories such as long-term memory (Figure 8–11).

We "fix" the vehicle's additional source of fuel by turning off the propane. Immediately, STFT begins to correct for what is now a lean condition by adding fuel (+%). Notice that in a short period of time both STFT and LTFT are the same number but in opposite directions. This is important to note, especially after you have repaired a vehicle. If STFT and LTFT are approximately equal but opposite, it indicates that the problem is solved. Eventually the STFT and LTFT will adjust down to zero.

Figure 8–12 shows a vehicle prior to repair. Notice that all correction seems to be LTFT oriented.

STFT is 0.0%, while LTFT is +14.8%. The PCM is compensating for a lean condition, which in this example was an open fuel injector on a 4-cylinder vehicle. The three functioning injectors were receiving a 14.8% increased signal. Once the repair is completed, what should the trim values do? With additional fuel available, the system should go rich and STFT should take away fuel. Figure 8–13 shows the results after the repair.

about:blank

46	639 (LG/P)	High Speed Cooling Fan Relay Control	79	911 (W/L
47	360 (BR/PK)	EGR Vacuum Regulator (EVR) Control		
48	11 (T/Y)	Tachometer	80	926 (LB/
49	–	NOT USED	81	925 (W/Y
50	–	NOT USED		
51	570 (BK/W)	Power Ground	82	–
52	851 (Y/R)	Ignition Coil B	83	264 (W/L
53	971 (PK/BK)	Shift Solenoid #3	84	970 (DG/
54	480 (P/Y)	Torque Converter Clutch (TCC) Solenoid	85	282 (DB/
55	1076 (DG/O)	Keep Alive Power (B+)		
56	101 (GY/Y)	(EVAP) Canister Purge Valve	86	347 (BK/
57	310 (Y/R)	Knock Sensor	87	392(R/LC
58	679 (GY/BK)	Vehicle Speed Sensor (VSS+) Input		
59	–	NOT USED	88	967 (LB/
60	74 (GY/LB)	Heated Oxygen Sensor #11	89	355 (GY/
		(HO2S) Input	90	351 (BR/
61	393 (P/LG)	Heated Oxygen Sensor #22 (HO2S)		
		Input (Catalyst Monitor)		
62	791 (R/PK)	Fuel Tank Pressure Sensor		
63	–	NOT USED		

Delmar/Cengage Learning

Figure 8-10 Pin 55 is identified as keep-alive memory.

Propane vs Trim

O2S	STFT	LTFT
450 mV	+/−0%	+/−0%
550 mV	Going down	+/−0%
450 mV	−20%	+/−0%
450 mV	−10%	−10%
450 mV	+/−0%	−20%
350 mV	Going up	−20%
450 mV	+20%	−20%

Delmar/Cengage Learning

Figure 8-11 When propane is turned off, a +20% correction to STFT occurs.

Notice that STFT and LTFT are very close to the same number but are opposite. This indicates that the vehicle has been repaired and the trim will adjust back to normal. First, the short-term fuel trim will correct the situation, and then the correction will be transferred over to the long term fuel trim. Again, take note that a vehicle that arrives in the shop with significant trim values is telling you something: additional fuel is either being added or subtracted. After you diagnose the vehicle and repair it, the trim values should be equal but opposite. After repair and driving, the trim values will eventually move back toward zero. This alone is a good reason to not clear diagnostic trouble codes during the

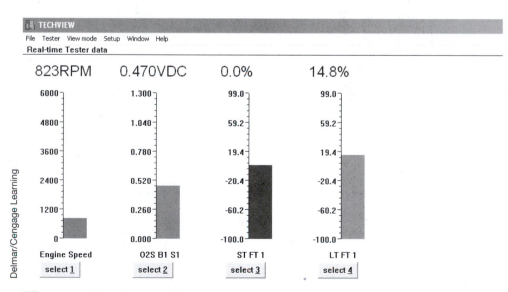

Delmar/Cengage Learning

Figure 8-12 14.8% trim is long term.

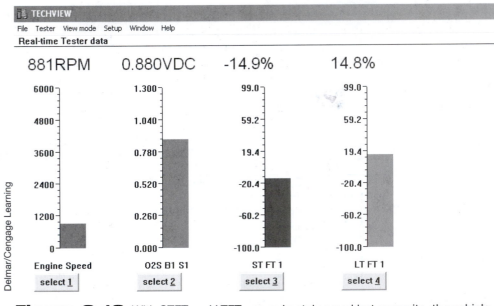

Figure 8-13 With STFT and LTFT approximately equal but opposite, the vehicle is fixed.

diagnosis and repair phase. You should also know at this point that the fuel monitor will adjust itself and turn off the check engine light once the conditions of freeze frame are met. This will allow most State inspections to occur even though the DTC is still in memory. The key is the light, not the code. If the light is on, the vehicle fails the emission test. If the light is off and sufficient monitors have been run, the vehicle passes.

Additional Fuel Trim Information There are times when fuel trim is a very good diagnostic aid. For example, if a mass air flow sensor (MAF) were to get contaminated, it would have an impact on fuel trim (Figure 8–14).

The sensor is responsible for detecting and measuring the quantity of air going into the engine. If the air filter is not regularly changed, or if an engine is run without an air filter, the sensor can become contaminated. It will now have difficulty accurately measuring the amount of air, resulting in the signal from the oxygen sensor indicating correction to the pulse width of the injectors is necessary. Most likely the MAF will overestimate the amount of air at idle. With this information the PCM will inject more fuel than is required. The O2S will indicate a rich condition, and the fuel trim will correct in the minus direction. At high speeds the situation is opposite; the MAF underestimates the amount of air, the PCM injects less fuel than is really needed, the

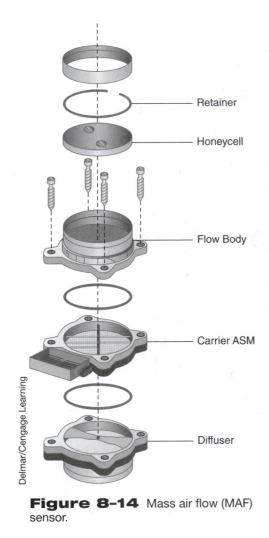

Figure 8-14 Mass air flow (MAF) sensor.

O2S signal indicates a lean condition, and the fuel trim adds fuel. A usual guideline is 13% at idle and high speeds, with the number being opposite and close to being equal. An engine with a contaminated MAF might show fuel trim numbers of −15% at idle and +14% at high speeds. The fact that they are approximately equal and yet opposite is the key to the contaminated MAF.

CONCLUSION

In this chapter, we examined the very important fuel systems monitor. We have looked at how the signal from the oxygen sensors is used to determine fuel adjustments. These fuel adjustments are the short-term fuel trim and the long-term fuel trim. Additionally we looked at the relationship between these two trim values under both a lean condition and a rich condition. We also discussed how the PCM retains the memory of STFT and LTFT.

REVIEW QUESTIONS

1. Technician A states that the fuel systems monitor is one of the continuous monitors. Technician B states that the oxygen sensors are one of the major inputs used in the fuel systems monitor. Who is correct?
   a. Technician A only
   b. Technician B only
   c. Both Technician A and B
   d. Neither Technician A nor B

2. A technician is looking at a DSO voltage pattern for a fuel injector. She should be able to see
   a. the applied voltage to the injector
   b. the applied ground to the injector
   c. the amount of time the injector is signaled to be open
   d. all of the above

3. A fuel injector voltage pattern is being observed as the engine is loaded down. Technician A states that the pulse width should increase with the additional engine load. Technician B states that the ground voltage will change as the engine load is increased. Who is correct?
   a. Technician A only
   b. Technician B only

   c. Both Technician A and B
   d. Neither Technician A nor B

4. Short-term fuel trim is changing, while long-term fuel trim is stationary. Technician A states that STFT is trying to get the engine back into fuel control. Technician B states that LTFT will begin changing once STFT is not moving. Who is correct?
   a. Technician A only
   b. Technician B only
   c. Both Technician A and B
   d. Neither Technician A nor B

5. O2S voltage rises. This should cause an instantaneous change of
   a. STFT should move down (negative)
   b. STFT should move up (positive)
   c. LTFT should move down (negative)
   d. LTFT should move up (positive)

6. One of the purposes of the keep-alive memory circuit is to
   a. allow certain accessories to be able to be run in the accessory position (key switch)
   b. store the results of short-term fuel trim
   c. maintain engine running data from the last start
   d. store the results of long-term fuel trim

7. After a vehicle is repaired and has run for a while, Technician A states that the STFT and LTFT will be equal but opposite. Technician B states that clearing the DTC is the best method for continuous monitor repair verification. Who is correct?
   a. Technician A only
   b. Technician B only
   c. Both Technician A and B
   d. Neither Technician A nor B

8. Two technicians are discussing the continuous monitors. Technician A states that the DTCs should always be cleared to turn off the check engine light. Technician B states that driving the freeze frame to rerun the monitor should be done. Who is correct?
   a. Technician A only
   b. Technician B only
   c. Both Technician A and B
   d. Neither Technician A nor B

9. A technician suspects that the mass air flow sensor is contaminated. This can be verified by
   a. looking for a contamination DTC
   b. checking the LTFT at low speeds and high speeds to see if it is opposite and approximately equal
   c. using a DSO on the oxygen sensor
   d. clearing the DTC to see if it comes back

10. Two technicians are discussing oxygen sensors. Technician A states that the O2S will generate a lower than 450 mV signal if the system detects a rich condition. Technician B states that the O2S signal will be used to adjust fuel delivery through the use of fuel trim. Who is correct?
    a. Technician A only
    b. Technician B only
    c. Both Technician A and B
    d. Neither Technician A nor B

COMPREHENSIVE COMPONENT MONITOR

OBJECTIVES

At the conclusion of this chapter you should be able to: ■ Recognize the difference between OBD I and the comprehensive component monitor ■ Understand the function and operation of the monitor ■ Understand how an OBD I or II system sets DTCs ■ Identify a DTC that is a rationality check ■ Follow the "drive the freeze frame" method of diagnosis or verification ■ Follow a wiring diagram or component layout and identify components that the monitor will check ■ Follow along the diagnosis and repair of circuit problems

INTRODUCTION

The last continuous monitor that we will look at is the comprehensive component monitor. It is most like the old OBD I system, which was designed to only look at input components. It allows "drive the freeze frame" to clear the check engine light as the other continuous monitors do and has the enabling criteria of start and run rather than a complicated driving sequence.

OBD I AS A STARTING POINT

The easiest way to understand this monitor is to go back to OBD I systems, which began with the computerization of the ignition and fuel systems and ended with the introduction of OBD II in 1996. OBD I systems had limited diagnostic capability, with as little as 17 DTCs in the 1980s. Let's look at a sensor and analyze the diagnostic function of OBD I. Figure 9–1 shows a throttle position (TP) sensor.

It is connected to the throttle body assembly and will move as the throttle is moved. It normally is held in place with two screws and might be adjustable, especially in the early days. Electrically, this sensor is one of a set of 5-volt sensors that are considered the basics of computerized engine controls. It is an input sensor designed to supply or deliver information to the PCM. The name gives an indication of the function; tell the PCM the position of the throttle. Figure 9–2 shows the electrical end of the sensor.

The three wires are fundamental to the function and diagnostic capability of the TP. Looking at the TP,

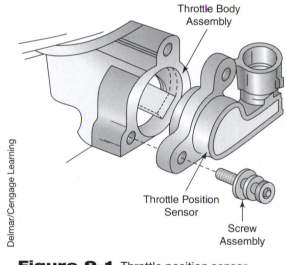

Delmar/Cengage Learning

Figure 9-1 Throttle position sensor attached to the throttle body.

A is a 5-volt feed (gray wire) source called *reference* on this GM vehicle. Reference is another way of saying controlled and constant. This 5-volt reference will be used to power up many sensors. It is terminal A on the sensor. A complete circuit through the TP is achieved with terminal B, a black wire that is connected to ground inside the PCM. Notice that inside the sensor there is a connection between A and B, schematically represented as a resistor. If you remember your basic electricity, a voltage drop occurs when current flows through resistance. With 5 volts on the top of the TP and ground (0 volts) on the bottom, the entire voltage will be dropped along the length of the resistance. If you were to measure this drop along the resistor, you would find it proportional to where you were measuring it. With the negative voltmeter probe on ground, if your voltmeter + probe is close to terminal A, you would find 5 V. If it were close to terminal B, you would find 0 V. Exactly in the middle of the resistance, you would find 2.5 V. Terminal C, which is labeled as TPS signal inside the PCM, is like the + voltmeter probe but moves with the throttle. Usually it never gets to either end of the resistor, but will deliver to the PCM a voltage from about .5 V to 4.5 V, with .5 being at closed throttle and 4.5 V being wide open throttle.

Now let's look at the diagnostic capability of OBD I on the TPS or TP. Figure 9–3 shows the same circuit but with a slight change.

Notice that the ground circuit from the TPS to the PCM has an open in it. Think your way through this one. It takes current flowing through resistance to have a voltage drop, and the open ground prevents any current flow. This means that because there is no

current flowing, there will be no voltage drop of the 5 volts. The TPS signal will show 5 volts, and when the PCM "sees" 5 volts on the signal line, it knows that the ground circuit must be open and will set a DTC for open ground circuit.

Figure 9–4 shows a similar problem but on the reference side.

You can probably figure out what will happen in this example, but let's walk through it. There is no reference 5 volts available to the TPS. As a result, the signal voltage will be zero and the PCM will set an open reference to the TPS DTC. This was the heart of the OBD I diagnostics: the ability to find and identify an open circuit on the + or – side of the circuit. Toward the end of the OBD I era the PCM was programmed to be able to see a nonfunctioning TPS, but most vehicles would generate only two or three DTCs per component. There was no ability to identify an out of range or inaccurate TPS. Probably one of the hallmarks of OBD II is the ability of the various monitors to be able to determine if a sensor is not making sense when it is compared to another. This is frequently called

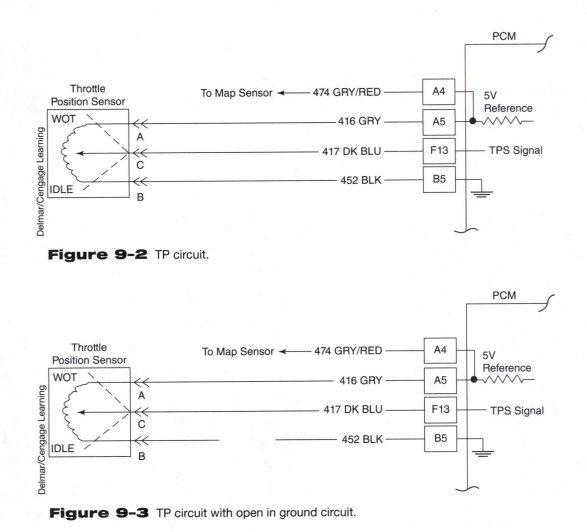

Figure 9-2 TP circuit.

Figure 9-3 TP circuit with open in ground circuit.

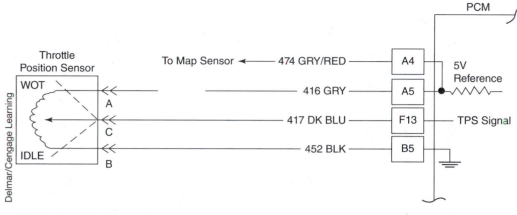

Figure 9-4 TP circuit with open in 5-volt reference.

Example P0121

- Title: Throttle sensor range/performance

Figure 9-5 A scanner screen with insufficient title information. Delmar/Cengage Learning

rationality checking. The PCM tries to determine if the signal makes sense or not.

A P0121 is a good example of this ability. Figure 9-5 shows a title for a P0121.

Notice that our scanner would show "Throttle sensor range/performance"; however, when we look up the definition of a P0121 in a manual or online, we see that actually the full title is "Throttle sensor range or performance does not agree with MAP." There is a significant difference in these two. One directs the technician to the TP as a problem, and the other directs the technician to either the MAP or the TP. This is a huge difference and illustrates the power of OBD II. OBD II has the ability to compare one signal to another and determine if there is a difference. Normally the signal of a TP and MAP follow one another, as Figure 9-6 shows.

The TP is on top and the MAP is on the bottom. Notice that the two traces follow one another very closely. As the throttle is opened rapidly (the top trace) to wide open throttle (WOT), the pressure in the manifold responds and increases to atmospheric (the flat top). After the throttle is closed, the pressure changes accordingly. What the OBD II system does with the continuous comprehensive component monitor is look at both sensors at the same time to determine if they respond the same.

It is very important to realize that the comprehensive component monitor will generate more codes

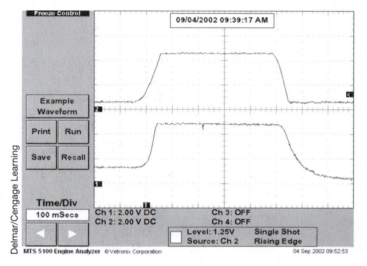

Figure 9-6 The map voltage (bottom trace) should follow the TP voltage (top).

than most other monitors. The EGR monitor might look at the ability of the system to flow exhaust gases, but the component monitor will look at each individual piece of the circuit including the sensors. Any circuit problems will be identified by this monitor. Figure 9-7 shows a common Ford EGR system utilizing a DPFE (differential pressure Feedback EGR). There are two very important electrical components, an EGR vacuum regulator and a DPFE, plus all of the circuit wiring connecting these two components to the PCM: power and ground. The comprehensive component monitor will check all of these components and be able to set many DTCs. The EGR monitor will only look at the flow, or the end result of the system. By examining the DTC set, the technician can zero in on a specific component.

P0400 through P040F (Figure 9-8) all relate to circuit, with only P0400, 401, and 402 specifically indicating

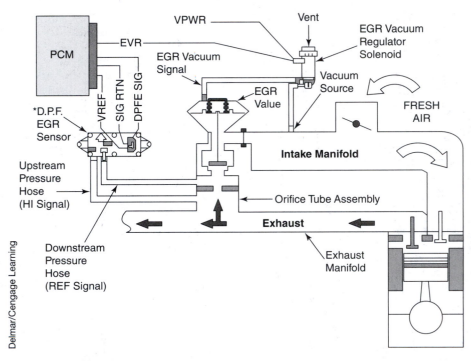

Delmar/Cengage Learning

Figure 9-7 Ford's DPFE system.

P0400	Exhaust Gas Recirculation A Flow
P0401	Exhaust Gas Recirculation A Flow Insufficient Detected
P0402	Exhaust Gas Recirculation A Flow Excessive Detected
P0403	Exhaust Gas Recirculation A Control Circuit
P0404	Exhaust Gas Recirculation A Control Circuit Range/Performance
P0405	Exhaust Gas Recirculation Sensor A Circuit Low
P0406	Exhaust Gas Recirculation Sensor A Circuit High
P0407	Exhaust Gas Recirculation Sensor B Circuit Low
P0408	Exhaust Gas Recirculation Sensor B Circuit High
P0409	Exhaust Gas Recirculation Sensor A Circuit
P040A	Exhaust Gas Recirculation Temperature Sensor A Circuit
P040B	Exhaust Gas Recirculation Temperature Sensor A Circuit Range/Performance
P040C	Exhaust Gas Recirculation Temperature Sensor A Circuit Low
P040D	Exhaust Gas Recirculation Temperature Sensor A Circuit High
P040E	Exhaust Gas Recirculation Temperature Sensor A Circuit Intermittent/Erratic
P040F	Exhaust Gas Recirculation Temperature Sensor A/B Correlation

Delmar/Cengage Learning

Figure 9-8 A single system may have many possible DTCs.

flow problems. The comprehensive component monitor will look at the components and the circuit.

Probably the most significant DTCs that come out of the comprehensive component monitor will be those that deal with circuits. This should direct the technician to perform various circuit tests, primarily looking for power/ground problems. When the DTC indicates circuit low, the ground is usually at fault.

OBD - REPAIR DIAGNOSTIC REPORT (RDR)

This Report contains information that will help a repair technician diagnose and repair your vehicle.

Plate: VIN: VIR: Make: NISS Model: MAXIMA Model Year: 1996	Test Time: 07/14/2005 11:15:43 Test No: 1 Vehicle Type: LDV, S Test Type: OBDII Station: 6 Lane: 3	MIL Commanded: ON Bulb Check: PASS Ready Status: READY

OBD TEST RESULT - FAIL

Your vehicle's computerized self-diagnostic system (OBD) registered the fault(s) listed below. This fault(s) is probably an indication of a malfunction of an emission component. However, multiple and/or seemingly unrelated faults may be an indication of an emission-related problem that occurred previously but upon further evaluation by the OBD system was determined to be only temporary. Therefore, proper diagnosis by a qualified technician is required to positively identify the source of any emission-related problem. The problems identified on this Repair Diagnostic Report must be corrected in order to pass the required OBDII test.

CODES	DESCRIPTION
P0150	02 SENSOR CIRCUIT (BANK 2 SENSOR 1)

FAILURE REASON: THE MALFUNCTION INDICATOR LIGHT (MIL) WAS COMMANDED ON FOR ONE OR MORE DIAGNOSTIC FAULT CODES LISTED ABOVE.

Delmar/Cengage Learning

Figure 9-9 The report from a vehicle with one DTC.

When the DTC indicates circuit high, the power or source is usually at fault.

Figure 9–9 shows a failed emission test for a 1996 Nissan Maxima. This vehicle was misdiagnosed by a shop that wound up replacing all of the oxygen sensors and the PCM before they gave up. The next shop recognized just what a P0150 was and that it involved the circuit not the sensor for bank 2 sensor 1. Remember that bank #2 is the bank on a V-type engine that does not have cylinder #1 in it and that sensor #1 is the front O2, the one primarily responsible for fuel control. Normally the O2Ss would be tested through the oxygen sensor monitor, one of the non-continuous monitors, which we will look at later in this text. The key to this P0150 is the fact that it involves the circuit. Circuits are diagnosed or monitored by the comprehensive component monitor. The repair on this vehicle was a simple one involving a broken wire between the sensor and the PCM.

A problem with some scanners is their inability to indicate the full description of the DTC due to the limited size of their display window. A "P0150 O2 Sensor circuit (bank 2 sensor 1)" could show as a "P0150 O2 sensor." This could be misinterpreted by a technician. The moral of the story is to look up the definition of the DTC if your scanner tends to shorten the description because of display window size. There are a variety of look-up websites available such as iATN. Once

you type in the DTC, the full description is displayed on your computer. Make a note of it and compare it to your scanner display window. They should be the same, but if they are not, use the full description to guide you to the problem. Going back to basics of checking the sensor for power, ground, and signal with a DSO is still the best avenue to take. The P0150 is a good example where replacing the sensor might not get at the circuit problem.

Because this monitor is almost exactly like the old OBD I system, the "clear the code to see if it comes back" diagnostic strategy does frequently work. Take the previous example of an open sensor circuit. If we clear the P0150 DTC, it will come back once the enabling criteria are met. With the enabling criteria basically being start and run, it is very easy to get the monitor to run and "see" if the problem still exists. There is, however, the problem of all of the monitors being cleared that we did not have in OBD I. There was only one monitor in OBD I, but there are many monitors in OBD II. Be very careful and aware of the consequences of clearing DTCs. Frequently, it makes more sense to fix the vehicle and drive the freeze frame, as we have discussed. If your repair was successful, the monitor will run the required three times and the MIL will go out. The vehicle is now ready for an emission test, should your State require one. If, instead, you decide to clear the code and see if it comes back, you will have to run a drive cycle to reset the monitors. This involves additional time and money, usually the customer's. Getting monitors to run can take as little as 15 minutes or as long as two hours. Additionally, some monitors will require an overnight soak of eight hours or so before they will run. This delay is unnecessary if you decide to drive the freeze frame. After the emission test, the DTC can be cleared and the customer will reset the monitors during usually normal driving over the next few days or so. Which method you use may also depend on the shop policy.

Let's look at another example of a DTC that was fired off of the comprehensive component monitor. A vehicle comes into the shop with a failed emission test. The State Vehicle Inspection Report is shown in Figure 9–10.

A P0102 is listed and the definition is Mass or Volume Air Flow Circuit Low Input. So this vehicle has a MAF circuit problem resulting in low input. This should be an easy repair once we realize the specifics of it. The first question in the technician's mind should be what monitor fired off this DTC? The answer is the comprehensive component monitor. The tech is now faced with some choices. One is to clear the code and see if it comes back, and frequently this will be done. But wait: clearing the DTC will also reset all monitors

to a not-ready status. What will be involved in getting the monitors to run? Figure 9–11 shows the steps.

You can see that although this vehicle is not too difficult, it still involves driving at 55 mph. Once 55 mph is achieved, an additional 20+ minutes will be required to drive the very specific sequence. Where the shop is located relative to a road that will allow 55 mph is the other factor that needs to be taken into consideration. Additionally, if the customer just filled the gas tank, the monitors will not run until the fuel level is below 75%. On the surface it does not look like clearing the DTC and seeing if it comes back is the best strategy. The first shop that looked at this vehicle charged the customer to replace the MAF, two O2Ss, and a diagnostic fee. The vehicle went back for the emission test and failed again. At this point the customer had hundreds of dollars in the vehicle and still could not get it to pass the required test.

Many States grade the technician's or shop's ability to fix emission repairs. A Blazer was brought to an "A" shop. Figure 9–12 is the final repair bill. Notice the underlined repair note: "Repaired Mass Air Flow Sensor Open Circuit." This ASE-certified and Emission "A" shop had no difficulty repairing the circuit. The DTC told the story or at least gave sufficient information to the shop to allow it to repair the real problem. The

OBD - REPAIR DIAGNOSTIC REPORT (RDR)

| View Test Details |

This Report contains information that will help a repair technician diagnose and repair your vehicle.

Plate: VIN: VIR: V013335575 Make: CHEV Model: BLAZER Model Year: 1997	Test Time: 12/12/2006 13:15:25 Test No: 1 Vehicle Type: LDT1, U Test Type: OBDII Station: 19 Lane: 3	MIL Commanded: ON Bulb Check: PASS Ready Status: READY

OBD TEST RESULT - FAIL

Your vehicle's computerized self-diagnostic system (OBD) registered the fault(s) listed below. This fault(s) is probably an indication of a malfunction of an emission component. However, multiple and/or seemingly unrelated faults may be an indication of an emission-related problem that occurred previously but upon further evaluation by the OBD system was determined to be only temporary. Therefore, proper diagnosis by a qualified technician is required to positively identify the source of any emission-related problem. The problems identified on this Repair Diagnostic Report must be corrected in order to pass the required OBDII test.

CODES	DESCRIPTION
P0102	MASS OR VOLUME AIR FLOW CIRCUIT LOW INPUT

FAILURE REASON: THE MALFUNCTION INDICATOR LIGHT (MIL) WAS COMMANDED ON FOR ONE OR MORE DIAGNOSTIC FAULT CODES LISTED ABOVE.

Delmar/Cengage Learning

Figure 9–10 A vehicle with one DTC.

Conditions	1. Baro >75 kPa 2. ECT <86 degrees F.; IAT <86 degrees F.; difference between ECT and IAT <7 degrees F. 3. Battery 10.5–16 volts; Fuel level 25%–75%
Step 1	Perform the I/M System Check. Failure to do so may result in difficulty in updating the monitor(s) status to YES.
Step 2	Turn ignition OFF for 5 minutes. Pre-program the scan tool with vehicle information before turning ignition ON. Start the engine and do not turn It OFF for the remainder of the test.
Step 3	Turn OFF all accessories and set parking brake. A/T should be in P and M/T should be in N. Idle the engine for 2 minutes.
Step 4	Accelerate at part throttle to 55 mph and maintain speed until engine reaches operating temperature. This could take 8–20 minutes depending on coolant temperature at start-up.
Step 5	Maintain speed of 55 mph for an additional 6–7 minutes.
Step 6	Reduce speed to 45 mph and maintain 45 mph for 1 minute.
Step 7	From 45 mph, perform 4 decelerations of 25 seconds each under the conditions in Step 8. Keep the speed above 25 mph and return to 45 mph for 15 seconds after each deceleration.
Step 8	Each deceleration period should be at closed throttle with NO brake application. NO clutch actuation and NO manual downshift.
Step 9	Accelerate at part throttle to 45–55 mph and maintain speed for 2 minutes.
Step 10	Decelerate to 0 mph and idle for 2 minutes with foot on brake pedal. A/T in D or M/T in N with Clutch depressed.
Step 11	Access the readiness status on the scan tool. Perform the individual drive cycle for any monitor that does not display YES (monitor is not ready).
Step 12	Check for DTCs. Any DTCs will require diagnosis and repair.
Step 13	Following repairs and clearing DTCs, perform Steps 1–12 again or perform steps for individual monitors that are not set.

Delmar/Cengage Learning

Figure 9-11 The enabling criteria may involve many steps with varying speeds and loads.

Invoice

Invoice No:	38963
Salesperson:	JLD
PO#	

Bar ID: 40724

ASE CERTIFIED

Customer: Address: City/State/Zip: HM: TDY:	Vehicle: 1997 CHEVROLET BLAZER 262cid 4.3L 0cc V6 FI GAS N W Mileage: 152,901 Color: GREEN LIC: 9645739 VIN:

Parts:	0.00		
Labor:	282.36	**Sub Total:**	**284.78**
Sublet Labor:	0.00	Sales Tax(9.000%)	0.22
Misc.:	0.00	**Total:** $	**285.00**
Additional Fees:	0.00	Tendered: (Cash)	285.00
Fixed Price Jobs:	0.00		
Shop Supplies:	2.42	**Change Due:** $	**0.00**

Part No.	Description	Qty/ Hours	Price Other	Labor	Parts	Disc.	Total
Labor	DIAGNOSTIC FEE						65.00
Labor	LABOR						217.36
SE	MALFUNCTION INDICATOR LIGHT MAY COME BACK ON FOR A NUMEROUS AMOUNT OF REASONS NOT RELATED TO THE REPAIRS PERFORMED AT THIS TIME.	1.00					

Comments:

CUSTOMERS COMPLAINT: FAIL EMISSION TEST

CODE PRESENT: P0102 MASS OR VOLUME AIR FLOW CIRCUIT LOW

REPAIRED MASS AIR FLOW SENSOR OPEN CIRCUIT.

90 DAYS WARRANTY

CLEARED PCM CODES

RUN OBD CYCLES

SET MONITORS TO READY

PASS EMISSION TEST

NO OTHER PROBLEM FOUND

TECH: ROD

Delmar/Cengage Learning

Figure 9-12 The DTC indicated a "circuit" that was the key to a successful repair.

shop cleared all codes and reran the drive cycle, re-setting the monitors to a ready status. The vehicle passed the emission test the same day.

Always read all of the information available to you when the check engine light is on. Frequently the scanner will give you only minimal information, with abbreviations for certain DTCs. Go to a look-up chart for OBD II DTCs to verify the full title and definition of the DTC.

CONCLUSION

In this chapter, we looked at the last continuous monitor—the comprehensive component monitor. We examined how the monitor generates DTCs based on individual components and circuits and saw how the full definition of the DTC may be invaluable. We also looked at how this monitor is like the OBD I monitor prior to 1996 because it generates data about open and short circuits. Driving the freeze frame was discussed, and the disadvantage of clearing the code to see if it comes back was analyzed. We also looked at the enabling criteria of all monitors as a discouragement to clearing the codes.

REVIEW QUESTIONS

1. Technician A states that the comprehensive component monitor is frequently compared to the OBD I system. Technician B states that the enabling criteria for the comprehensive component monitor are "start and run." Who is correct?

   a. Technician A only

   b. Technician B only

   c. Both Technician A and B

   d. Neither Technician A nor B

2. The comprehensive component monitor

   a. is one of the continuous monitors

   b. is used to test the electrical circuits of most components

   c. responds well to a "clear the code and see if it comes back" type of diagnosis

   d. all of the above

3. Two technicians are discussing component DTCs. Technician A states that components are tested during a specific drive cycle. Technician B states that component circuits are tested by the comprehensive component monitor. Who is correct?

   a. Technician A only

   b. Technician B only

   c. Both Technician A and B

   d. Neither Technician A nor B

4. A DTC indicates that throttle sensor range or performance does not agree with MAP. This DTC is an example of one that tests

   a. the sensor's applied power

   b. the sensor's ground

   c. the rationality between two sensors

   d. none of the above

5. Technician A states that driving the freeze frame involves having the same conditions that were present during the setting of the DTC. Technician B states that all monitors will run while driving the freeze frame. Who is correct?

   a. Technician A only

   b. Technician B only

   c. Both Technician A and B

   d. Neither Technician A nor B

6. An EGR circuit is being diagnosed that has three electrical components. Technician A states that the circuit will be tested during the EGR monitor. Technician B states that the three electrical components will be tested during the running of the comprehensive component monitor. Who is correct?

   a. Technician A only

   b. Technician B only

   c. Both Technician A and B

   d. Neither Technician A nor B

7. A DTC has the word *circuit* in its definition. This indicates

   a. a bad sensor

   b. the PCM cannot "see" the circuit

   c. the system is not functional

   d. there is a problem with power, ground, or the signal relative to the circuit

8. Two technicians are discussing the continuous monitors. Technician A states that specific enabling criteria are required to get the monitors to run. Technician B states that the three

monitors are most like the old OBD I system. Who is correct?

a. Technician A only

b. Technician B only

c. Both Technician A and B

d. Neither Technician A nor B

9. An EGR flow DTC is present with an EGR circuit DTC. Technician A states it is necessary to begin with the EGR flow DTC. Technician B states it is necessary to begin with the circuit DTC. Who is correct?

a. Technician A only

b. Technician B only

c. Both Technician A and B

d. Neither Technician A nor B

10. A circuit DTC is set. What should a technician test first?

a. power to the circuit

b. ground to the circuit

c. signal from the sensor

d. all of the above should be tested in any order

OXYGEN SENSOR MONITOR

OBJECTIVES

At the conclusion of this chapter you should be able to: ■ Identify the position of an O2S based on its name ■ Identify a rich or lean condition off of a DSO pattern ■ Recognize the signal from a B1S2 O2S as an indication of CAT efficiency ■ Run the O2S monitor on a variety of vehicles ■ Recognize the various DTCs that the oxygen sensor monitor may set

INTRODUCTION

The oxygen sensor monitor is responsible for the testing and verification of one of the most important sensors within the OBD II system. It involves two separate monitors: One for the heater and one for sensor operation. In this chapter we will look at how both of the monitors function, their enabling criteria, and the DTCs that may set.

THE OXYGEN SENSOR HEATER

A part of the overall testing of the oxygen sensor is the testing of the heater. The heater is an integral part of the function of the sensor. It allows the signal to be useful to the PCM in a very short period of time, after a cold start. The circuitry that brings current to the heater is simple and normally involves the use of the PCM. Figure 10–1 Because the PCM is involved

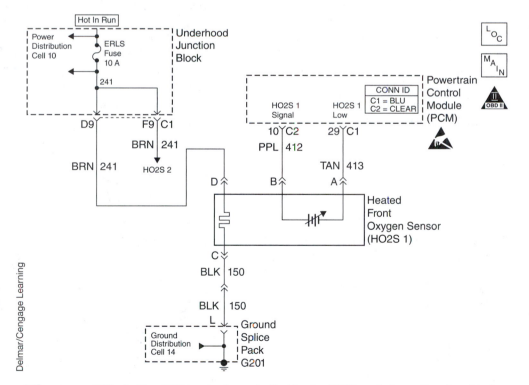

Figure 10-1 The O2S heater is controlled by the PCM and allows the signal to be accurate even at low temperatures

Delmar/Cengage Learning

directly in heater current flow, it is easy for the monitor to determine if the heater is functional. When the monitor runs, the PCM looks to see that the circuit draws current. If it draws current, is will usually pass the monitor. If there is no current flow, the monitor runs and sets a DTC. This is usually the first monitor to run and perhaps the most important since it allows the oxygen sensor to run and be functional shortly after start up. Once the oxygen sensor is hot, many of the manufacturers turn off the circuit since heater current is no longer required.

WHAT IS THE OXYGEN SENSOR MONITOR?

What the oxygen sensor (O2S) is responsible for is a good place to start when looking at this monitor. The basic air/fuel mixture is determined by the PCM and delivered as a pulse width to the fuel injectors. If the vehicle needs are for more fuel, the pulse width is increased. If the vehicle needs less fuel, the pulse width is decreased. We looked at this in Chapter 8, where we examined the fuel systems monitor. The O2Ss begin their work as the burn takes place within the combustion chamber. There are basically two different jobs done by the O2Ss based on their position. Figure 10–2 shows the positions that are possible.

Let's review a bit prior to looking at the monitor. Bank 1 has cylinder number 1, and bank 2 is the opposite bank. An in-line engine only has bank 1. The sensors labeled as number 1 are the ones used for fuel adjustment, since they "see" the results of the burn immediately. So in Figure 10–2 B1S1 and B2S1 are the front O2Ss, and their function will be to indicate to the PCM the characteristics of the burn. The voltage from the sensor is proportional to the level of oxygen remaining after the burn. As the oxygen rises, the

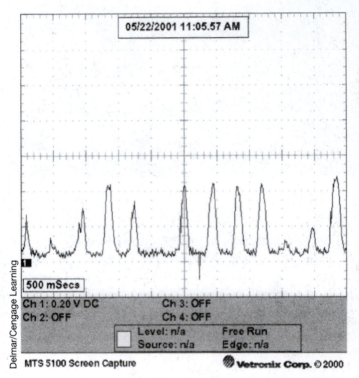

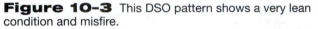

Figure 10–3 This DSO pattern shows a very lean condition and misfire.

system is running lean and the average voltage is low. Figure 10–3 shows a very lean running vehicle. The average voltage should be around .450 V (450 mV).

Notice that on the DSO pattern shown here the voltage barely rises above 0 V. The peaks do not even make it to 450 mV, much less the average. This is a captured waveform from a vehicle in need of a new fuel pump. This sensor's output is running between 50 mV and 400 mV, with an average of around 100 mV. The pump's volume has been reduced to the point of lean misfire. Lower than 450 mV average indicates a lean condition.

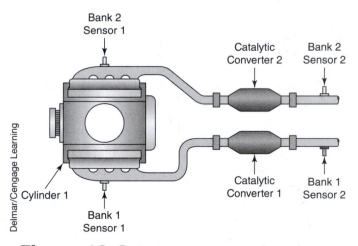

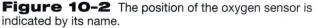

Figure 10-2 The position of the oxygen sensor is indicated by its name.

The opposite is the rich running vehicle shown in Figure 10-4. Notice that the average voltage now is well above the 450 mV mark. This sensor is running between 800 mV and 1 Volt and is indicating a very rich running system. The B1S1 and B2S1 signals are used to adjust the mixture through the use of fuel trim, which we covered in Chapter 8. There are no more important sensors to the running of the vehicle or to the emissions produced than the S1 oxygen sensors. Their signal is used to adjust the mixture to keep the emissions within acceptable standards.

The S2 sensors have a different function. Their signal is used to determine catalytic convertor efficiency. A comparison between the S1s and the S2s is done to indicate the efficiency. Figure 10-5 shows the signals coming from four O2Ss on a dual-CAT vehicle. Notice that both S2s are flat.

The oxygen level coming out of the CAT is really not changing. This is because the CAT is doing its job well. The oxygen levels going in are varying (the S1s). The mixture varies going into the CAT but is flat coming out. This indicates a very high level of efficiency within the CAT.

Let's look at another example but for a CAT that is not functioning at normal levels. Figure 10-6 shows a single CAT vehicle (4 cylinder).

The top trace is the S1 sensor, which looks normal. However, the S2 sensor (the bottom trace) shows lots of activity, indicating less than normal efficiency within the CAT. It is the ratio between the front O2S and the rear O2S that will indicate CAT efficiency. Without functioning O2Ss, CAT efficiency cannot be calculated correctly. This is an example of a monitored sensor being used in the monitoring of another system. It is one of the reasons why the O2Ss are extremely important and need to be monitored for function and accuracy. O2S signals are frequently used to determine how well the CAT, EVAP, and EGR system are functioning. The signal from the O2S must be an accurate indicator of what is going on.

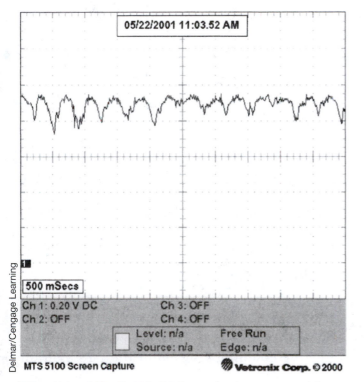

Figure 10-4 This DSO pattern shows a very rich condition and misfire.

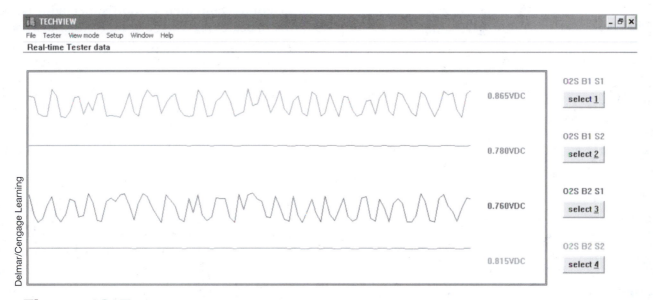

Figure 10-5 This scan graph shows good activity on all O2Ss.

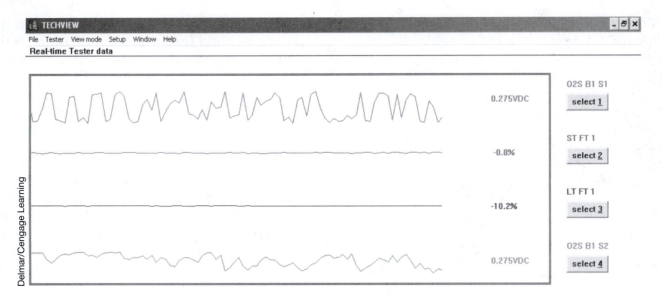

Figure 10-6 B1S2 shows activity indicating possible bad CAT.

HOW DOES THE MONITOR FUNCTION?

The O2S monitor functions under varying conditions and follows the propane method of testing an O2S, shown later in this text. As the driver removes his foot from the gas and the vehicle begins to decell, the PCM turns on the fuel injectors and watches the O2S response. Under decell conditions the system is running very lean and the PCM should see a lean signal. By turning on the fuel injectors for a split second, the system should go rich. The DSO pattern in Figure 10-7 shows an O2S with enrichment. Notice that the pattern flatlines rich. The transition from lean to rich is also observed to determine how fast the O2S is capable of reacting. In this way the three important characteristics of a functioning O2S are tested: lean voltage, rich voltage, and speed.

In our example the rich voltage is about 900 mV, the lean voltage is around 100 mV, and the lean-to-rich voltage transition is just about instantaneous.

ENABLING CRITERIA

The enabling criteria for testing the O2Ss are usually not complicated but still involves various speeds and loads. Let's look at a common vehicle and examine the enabling criteria for the O2S monitor (Figure 10-8).

Notice that on this Ford Escape, the fuel tank fill level is a requirement and three-quarters full is suggested. Additionally, the comment is made to operate the throttle smoothly or more time may be required. Some idle is required and then 40 mph until engine temp is above 170 degrees F. When a 40 mph cruise of five minutes duration after reaching 170 F is achieved, the monitor should run. This is not a complicated monitor and should run easily. As it runs, it may generate a variety of DTCs that we will look at later. Let's take another vehicle, one that is not quite so simple (Figure 10-9).

This 1997 Isuzu does not have a separate O2S monitor drive cycle, so the "all monitor" cycle must be followed, and it has 13 steps! Notice that within the steps there are very specific speeds, time, and acceleration/deceleration rates. Some are difficult to attain, based on where the repair shop is located. The question that must be answered by the technician is: Can I drive the vehicle this specifically around the shop area and still accomplish the drive cycle? Once the technician is into the cycle, it must be completed. This is done to get all the monitors set, and it is true

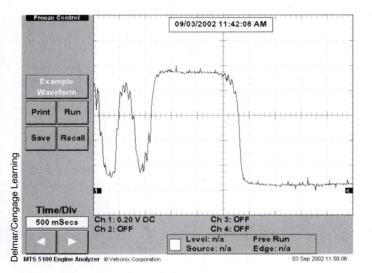

Figure 10-7 Both the rich and lean voltage indicate an accurate O2S.

OBD II Drive Cycles		Ford Escape
All Monitors		**2005**
Conditions	1. Fuel tank should be 1/2–3/4 full; 3/4 full is preferable. 2. Operating the throttle smoothly during cruise or acceleration will minimize the time required for monitor completion.	
Step 1	Connect a scan tool to the data link connector. Turn the key ON with the engine OFF. Cycle the key OFF, then ON. Clear all DTCs and reset the PCM.	
Step 2	Monitor the following PIDs: ECT, EVAPDC, FLI (if available) and TP MODE. Start the vehicle without returning to key OFF.	
Step 3	Idle the engine for 15 seconds, then drive at 40 mph until the ECT is at least 170 degrees F.	
Step 4	Cruise at 40 mph for at least 5 minutes.	
Step 5	Check monitor status on scan tool.	
Step 6	If any non-continuous monitor has not completed. Check for pending codes and make any necessary repairs. Re-run any incomplete monitors.	

Figure 10-8 The enabling criteria for a Ford Escape's O2S.

OBD II Drive Cycles		Isuzu Rodeo
All Monitors		**1997**
Notes	1. This drive cycle is designed to simulate highway driving. 2. When instructed, vary speed smoothly. 3. Also perform the All Monitors - Urban (City) Drive Cycle twice.	
Conditions	1. Cold start.	
Step 1	Idle 20 seconds. Accelerate gradually and drive at 20–25 mph for 1 minute. Vary speed.	
Step 2	Drive at 25–31 mph for 35 seconds. Decelerate to 0 mph in 10 seconds. Idle 40 seconds.	
Step 3	Accelerate moderately. Drive at 20–25 mph for 20 seconds. Increase speed to 40–55 for 85 seconds. Then decelerate to 0 over 50 seconds. Idle 15 seconds.	
Step 4	Gradually increase speed to 36 mph in 35 seconds. Decelerate to 0 in 15 seconds. Idle 10 seconds.	
Step 5	Accelerate to 30 mph and decelerate to 0 over 25 second period. Idle 20 seconds.	
Step 6	Accelerate to 36 mph in 20 seconds. Drive at 35 mph for 20 seconds, Decelerate to 0 in 15 seconds. Idle 5 seconds.	
Step 7	Accelerate to 26 mph and decelerate to 0 in 40 seconds. Idle 15 seconds.	
Step 8	Accelerate to 27 mph in 40 seconds. Decelerate to 0 in 8 seconds. Idle 25 seconds.	
Step 9	Accelerate to 26 mph and decelerate to 0 in 35 seconds. Idle 15 seconds.	
Step 10	Drive in stop-and-go traffic for 1 minute, reaching 25–30 mph twice, with no complete stops.	
Step 11	Drive at 20–30 mph for 2 minutes and stop. Vary speed. Drive at 20–28 mph for 2–1/2 minutes at varying speeds. Stop. Idle 30 seconds.	
Step 12	Accelerate to 28 mph and back to 0 in 50 seconds. Accelerate to 20 mph in 10 seconds. Drive at 20–27 mph for 20 seconds and decelerate to 0 in 10 seconds. Idle 15 seconds.	
Step 13	Accelerate to 23 mph and back to 0 in 20 seconds. Idle 10 seconds. Accelerate to 22 mph and back to 0 in 45 seconds. Idle 10 seconds.	

Figure 10-9 Sometimes the enabling criteria will allow one drive cycle to run all monitors.

that the O2S monitor might set somewhere during the cycle and not require all 13 steps. But the fact is some monitors are very difficult to run, especially since they will have to run twice to change the status screen from not run to run (Figure 10–10).

These screens are the monitor status screens, and they show that the O2S monitor has run. Remember this does not indicate that the system has been passed, only that the monitors have run the required two times.

Figure 10-10 This scanner indicates that all monitors have run.

DTCS ASSOCIATED WITH THE O2S MONITOR

Once the O2S monitor runs, it is capable of setting or identifying various problems. The problems will usually result in a variety of DTCs. Figure 10–11 shows a typical list of ten possible DTCs for the front O2S.

Additionally, slow response has its own set of DTCs as shown in Figure 10–12.

Notice that for this monitor to detect slow response, there is an expensive list of sensors that must first be determined to be "OK." These are for the front O2Ss, either B1S1 or B2S1.

HO2S "Lack of Switching" Operation:	
DTCs	P1131 or P2195 - Lack of switching, sensor indicates lean, Bank 1
	P1132 or P2196 - Lack of switching, sensor indicates rich, Bank 1
	P0132 Over voltage, Bank 1
	P1151 or P2197 - Lack of switching, sensor indicates lean, Bank 2
	P1152 or P2198 - Lack of switching, sensor indicates rich, Bank 2
	P0152 Over voltage, Bank 2
Monitor execution	Continuous, from startup and while in closed loop fuel
Monitor Sequence	None
Sensors OK	TP, MAF, ECT, IAT, FTP
Monitoring Duration	30 to 60 seconds to register a malfunction

Figure 10-11 O2S lack of switching can result in numerous DTCs.

HO2S Response Rate Operation:	
DTCs	P0133 (slow response Bank 1)
	P0153 (slow response Bank 2)
Monitor execution	Once per driving cycle
Monitor Sequence	None
Sensors OK	ECT, IAT, MAF, VSS, CKP, TP, CMP, no misfire DTCs, FRP
Monitoring Duration	4 seconds

Figure 10-12 Monitor specifics related to slow response DTCs.

Rear HO2S Check Operation:	
DTCs Sensor 2	P0136 HO2S12 No activity or
	P2270 HO2S12 Signal Stuck Lean
	P2271 HO2S12 Signal Stuck Rich
	P0138 HO2S12 Over voltage
	P0156 HO2S22 No activity or
	P2272 HO2S22 Signal Stuck Lean
	P2273 HO2S22 Signal Stuck Rich
	P0158 HO2S22 Over voltage
DTCs Sensor 3	P2274 HO2S13 Signal Stuck Lean
	P2275 HO2S13 Signal Stuck Rich
	P0144 HO2S13 Over voltage
	P2276 HO2S23 Signal Stuck Lean
	P2277 HO2S23 Signal Stuck Rich
	P0164 HO2S23 Over voltage
Monitor execution	Once per driving cycle for activity test, continuous for over voltage test
Monitor Sequence	None
Sensors OK	
Monitoring Duration	Continuous until monitor completed

Delmar/Cengage Learning

Figure 10-13 The after-the-CAT DTCs that are possible.

If we take a look at the rear O2Ss, the ones responsible for CAT efficiency, we see a different set of DTCs (Figure 10–13).

Notice that there are some P0s and some P2s available. The P2s are manufacturer specific, in this case Ford. The PCM wants to see a slowly moving voltage from the rear O2Ss. This "trending" indicates that the CAT is functioning within design limits.

CONCLUSION

In this chapter, we looked at probably the most important monitor, the oxygen sensor monitor. This monitor is critical because other systems will rely on the oxygen sensor for their function. The CAT monitor is a good example; determining the efficiency of the CAT involves the O2S's signal. With as many as six O2Ss on a vehicle, it is vital to the functioning of the OBD II system. The O2S monitor is also one of the first monitors to typically run. The O2S monitor's enabling criteria vary among manufacturers and can be very simple or very complicated. We also looked at the various DTCs that can be set by running the O2S monitor.

REVIEW QUESTIONS

1. Two technicians are discussing the oxygen sensor monitor. Technician A states that the O2S monitor should run prior to other monitors that will use O2S sensor data. Technician B states that the CAT monitor will use the S1 and S2 (or S3) information to determine CAT efficiency. Who is correct?

   a. Technician A only

   b. Technician B only

   c. Both Technician A and B

   d. Neither Technician A nor B

2. Which O2S is responsible for fuel trim information?

   a. B2S2

   b. B1S2

   c. B1S1

   d. B2S3

3. An O2S signal is averaging around 100 mV. Technician A states that this indicates the vehicle is running rich. Technician B states that

this should result in a negative fuel trim. Who is correct?

a. Technician A only

b. Technician B only

c. Both Technician A and B

d. Neither Technician A nor B

4. A comparison between B1S1 and B1S2 shows that both have lots of activity. This indicates

a. the CAT is functioning efficiently

b. the O2S is producing a false signal

c. the CAT is NOT functioning efficiently

d. the O2S monitor should set O2S DTCs

5. Technician A states that the O2S monitor will add and remove fuel, looking at response time and voltage. Technician B states that the monitor will typically function during engine idle conditions. Who is correct?

a. Technician A only

b. Technician B only

c. Both Technician A and B

d. Neither Technician A nor B

6. The enabling criteria for the O2S are observed and show that the engine should be at 170 degrees F minimum. If the O2S monitor were to run at 50 degrees F, what might be the result?

a. False DTCs might be set.

b. Other monitors (CAT, EVAP, EGR., etc.) might not run to completion.

c. The O2S monitor might not run the required second time.

d. all of the above

7. A vehicle does not show O2S monitor enabling criteria, but does show an all-monitor sequence. Technician A states that the vehicle does not have

an O2S monitor. Technician B states that the O2S monitor should run during the all-monitor drive cycle. Who is correct?

a. Technician A only

b. Technician B only

c. Both Technician A and B

d. Neither Technician A nor B

8. The readiness screen on the scanner shows all monitors have run. Technician A states that this means all monitors have passed all systems with no DTCs. Technician B states that this indicates the vehicle is ready for the enabling criteria to be run. Who is correct?

a. Technician A only

b. Technician B only

c. Both Technician A and B

d. Neither Technician A nor B

9. A lack of switching DTC shows for B1S1 O2S. This indicates

a. the fuel trim information might not be accurate

b. the level of exhaust coming out of the CAT will not be accurately measured

c. the vehicle must be experiencing a misfire

d. all of the above

10. Technician A states that the B1S2 sensor is showing a slowly moving (changing) voltage, indicating it needs to be replaced. Technician B states that the B1S1 sensor should have oscillating activity or it might be bad. Who is correct?

a. Technician A only

b. Technician B only

c. Both Technician A and B

d. Neither Technician A nor B

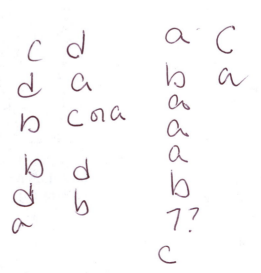

CAT MONITOR

OBJECTIVES

At the conclusion of this chapter you should be able to: ■ Explain why a CAT is important ■ Understand the basics of the CAT monitor ■ Utilize the O2S's signals to analyze the function of the CAT ■ Analyze the B1S1 and B1S2 O2S signals for ratio ■ Be able to run the enabling criteria for the CAT monitor

INTRODUCTION

Now that we have examined the oxygen sensor monitor, it is time to begin looking at one of the systems that will use the O2S signals. It is very important that you realize the function of the catalytic converter (CAT) is number one in the OBD II system. A functioning CAT is responsible for the passing of the FTP (Federal Test Procedure). Without the CAT, the specifications of the FTP cannot be met. In this chapter we will examine how the CAT monitor functions.

WHAT IS THE CAT MONITOR?

To understand the CAT monitor requires first a basic working knowledge of the catalytic converter. All vehicles produce, during the combustion burn, byproducts that are considered pollution. Gasoline is primarily made up of hydrocarbons, which, when mixed with sufficient oxygen and burned, will produce power. Complete combustion, which rarely occurs, produces carbon dioxide and water. However, it is the incomplete combustion that we are primarily concerned with. Some hydrocarbon remains unburned, and some of the burn produces carbon monoxide. The function of the CAT is to take the hydrocarbon and carbon monoxide and convert them into carbon dioxide and water. The third pollutant is nitrogen oxide or NO_x. Air is close to 80% nitrogen, and although it is the 20% of oxygen that will support the combustion process, it is the nitrogen that will produce the NO_x. When internal cylinder temperatures get to around 2,500 degrees F, the nitrogen

will begin to join with some of the oxygen, forming the NO_x.

Inside the stainless steel shell of the CAT is a catalyst (Figure 11–1). The exhaust will flow through the catalyst; and through a very high heat chemical reaction, the three pollutants—HC, CO, and NO_x—will be converted into CO_2 and water. You may read about two-way or three-way converters. A two-way converter reduces HC and CO only and is generally not found on vehicles produced after the late 1980s. All OBD II vehicles use three-way converters, with the third gas being NO_x. And this is where the problem occurs. The chemistry to convert NO_x is most effective with a rich mixture, while the chemistry to convert HC and CO is most effective with a lean mixture (Figure 11–2).

Notice that at richer air/fuel (A/F) ratios, NO_x conversion is at its highest and HC and CO are at their lowest. As the A/F ratio leans out, HC and CO conversion increases and NO_x drops off rapidly. The small window of "ideal" A/F ratio has to be maintained or the modern three-way convertor will not meet specification. This is why the function of the front oxygen sensors is so important—they indirectly control what goes into the convertor.

If the A/F ratio is correct, the mixture will alternate between rich and lean. When the mixture is rich, NO_x conversion is high; and when the system is lean, HC and CO conversion is high. In this way a single converter can control all three pollutants. The key to the system is controlling the air/fuel ratio very tightly within an ideal range called *stoichiometric*

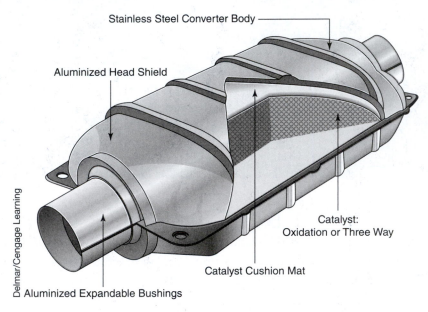

Stainless Steel Converter Body

Aluminized Head Shield

Delmar/Cengage Learning

Catalyst:
Oxidation or Three Way

Catalyst Cushion Mat

Aluminized Expandable Bushings

Figure 11-1 The inside of a three-way catalytic converter.

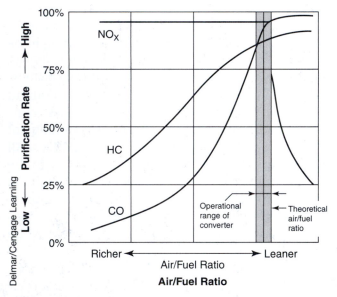

Figure 11-2 Air/fuel mixture has a direct impact on converter efficiency.

or *lambda* (the dark band on the Figure 11–2 chart) and allowing it to oscillate slightly between rich and lean. So you can see why the oxygen sensor monitor and functioning O2Ss are important. Without a tight control of the air/fuel ratio, the CAT cannot do its job. It is normal to look at the list of monitors run and see that the CAT monitor has not run, if the O2S monitor has not run or has generated DTCs. The PCM has been programmed to look at the converter only after verifying that the O2Ss are functional.

You will notice that frequently the enabling criteria for the CAT monitor include a statement regarding engine temperature and also the use of high-speed driving. This is because the CAT will not function below about 550 degrees and will not be fully efficient until over 750 degrees (Figure 11–3).

So one of the functions of OBD II is to test the CAT against predetermined standards and either pass or

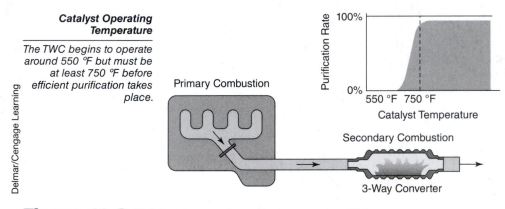

Catalyst Operating Temperature

The TWC begins to operate around 550 °F but must be at least 750 °F before efficient purification takes place.

Primary Combustion

Secondary Combustion

3-Way Converter

Figure 11-3 Catalyst temperature affects converter efficiency.

fail. By now you know that this testing of the CAT will be done by the CAT monitor. The basics are that the CAT monitor will look at the signals coming off of the front O2Ss and compare them to the signals coming off of the rear O2Ss. Within the functioning OBD II CAT, the three pollutants will be controlled but only if the CAT is capable. The CAT remains one of the most important, if not the most important, emission component. Its function and ability to control the three pollutants is considered one of the primary functions of the OBD II system. It then follows that the CAT must be monitored to determine its efficiency.

HOW DOES THE CAT MONITOR FUNCTION?

It is important to understand that the monitoring of the CAT will be done by the placement of two oxygen sensors: one pre-CAT and one post-CAT, as Figure 11–4 shows. The pre-CAT O2S is generally also the one that will determine the fuel trim adjustments to the calculated injector pulse width. So the pre-CAT or front O2S has two functions: determining the air/fuel ratio adjustments and indicating to the

CAT monitor what is going into the CAT. It is most frequently referred to as the S1 sensor. The post-CAT sensor is the S2 sensor. The sensor is usually referred to as the B1S1. The B1 stands for bank 1, which contains cylinder number 1. B2S1 is the pre-CAT sensor for bank 2, which does not contain cylinder 1. You will see both B1 and B2 on engines that are a V-type configuration (V-6 or V-8 usually). A 4-cylinder or 6-cylinder in-line engine only contains one bank, and so you will only see B1S1. This configuration is shown in Figure 11–4.

There are some engines that use two CATs, one for each bank, so you may see four O2Ss in use: B1S1 and B2S1 for input to the CATs and B1S2 and B2S2 for after- or post-CAT functions. Remember that the most important function of the S2s is to determine what is coming out of the CAT. Figure 11–5 shows multiple O2S applications for V type engines.

Figure 11–6 shows a diagram of what the signals should look like for a vehicle with a functioning CAT.

Figure 11–7 shows the same style of diagram with a non-functioning CAT.

The more activity that shows on the post-O2S, the less effective the CAT or CATs are. If we look at actual scan data taken off of a vehicle, this might make more

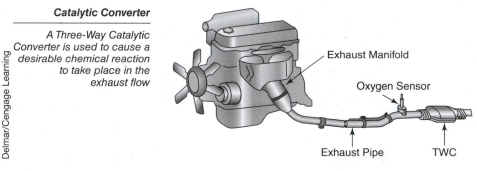

Figure 11–4 All OBD II vehicles use three-way catalyst.

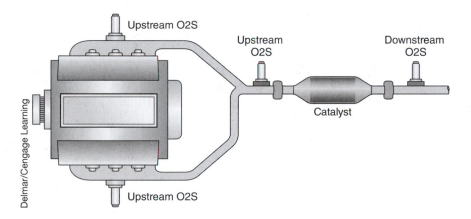

Figure 11–5 A V-6 engine with four oxygen sensors.

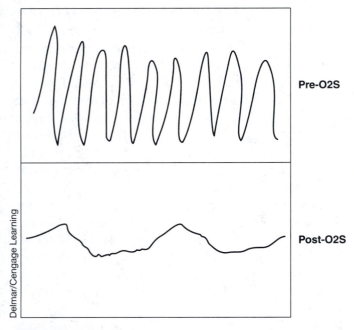

Figure 11-6 The lack of activity on the post-O2S indicates a good CAT.

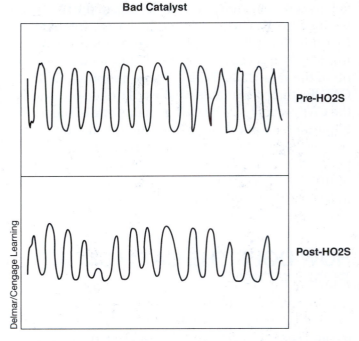

Figure 11-7 A bad catalyst will cause the post-O2S to have activity similar to the pre-O2S.

sense. Figure 11–8 is a scan off of a dual-CAT Lincoln with good CATs. The vehicle is cruising under steady gas pedal at around 55 mph.

Notice that B1S1 and B2S1 show lots of activity from the input O2Ss. This vehicle is in fuel control with no MIL on or DTCs. It is functioning correctly. Notice that the B1S2 and the B2S2 patterns are basically a flat line. If the CAT is functioning, there will not be swings in O2S voltage like the pre-O2Ss show. Instead, there will be a flat line slowly moving up and down over long periods of time. The CAT action is to use the leftover oxygen when the system is rich and store oxygen when

it is lean. Because it is stored and then used, the oxygen does not swing up and down in a functioning CAT. Figure 11–9 shows the opposite situation.

This is a scan from a single CAT vehicle. The top trace is from B1S1, the pre-CAT O2S, and the bottom scan is from the B1S2, the post-CAT O2S. This vehicle has the MIL on and has stored a P0420, a catalyst efficiency DTC. It is easy to see that the post-CAT O2S (B1S2) has lots of oxygen swings. Apparently this vehicle's CAT is not storing and then using the oxygen that is coming in with the exhaust stream.

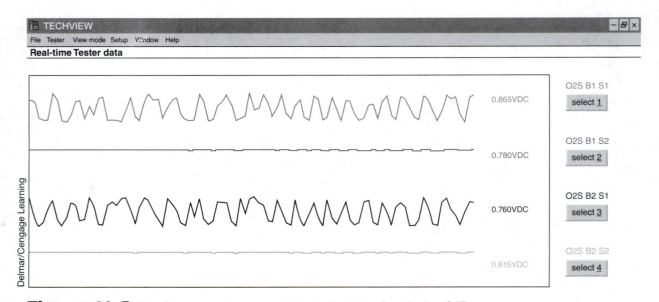

Figure 11-8 The S2 sensors have no activity, indicating a functioning CAT.

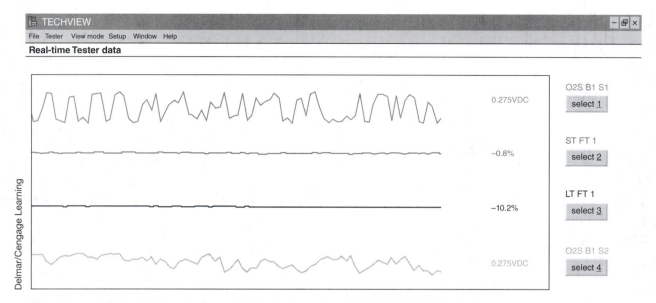

Figure 11-9 B1S2 indicates a CAT that will result in a P0420 DTC (CAT efficiency below threshold).

When the monitor ran, it saw substantial activity on the post-CAT O2S and was able to determine that the CAT was not up to the standards required and turned on the MIL. Remember from previous chapters that an OBD II vehicle turns on the MIL and sets a DTC if the vehicle could be producing 1.5 times the FTP standards. The vehicle shown in Figure 11-8 has run the monitor, determined that the CAT is not efficient enough, figured that the emissions out the tailpipe are probably 1.5 times the FTP, and turned on the MIL. The other thing this figure shows is the importance of the post-CAT O2S. Without the signal, the PCM has no way of calculating the efficiency of the CAT.

DTCS ASSOCIATED WITH THE CAT MONITOR

The most obvious DTC is the "popular" P0420—catalyst efficiency bank 1. There also is an extensive list of CAT codes displayed in Figure 11-10.

The most frequent CAT codes are those that deal with efficiency: P0420 and P0430, with the 430 being for bank 2. Few vehicles have dual CATs, so a P0430 is not common. But a P0420 is usually the number one DTC in States where emission testing is mandatory. Notice that some of the DTCs are for "heated catalyst." Currently there are no vehicles that use a heated catalyst, but when the list was made, it was thought that the future might require a series of DTCs for heated systems.

Remember that these DTCs are available only if the monitor runs, so the enabling criteria become the all-important factor in setting a DTC and turning on the MIL.

- P0420 Catalyst System Efficiency Below Threshold (Bank 1)
- P0421 Warm Up Catalyst Efficiency Below Threshold (Bank 1)
- P0422 Main Catalyst Efficiency Below Threshold (Bank 1)
- P0423 Heated Catalyst Efficiency Below Threshold (Bank 1)
- P0424 Heated Catalyst Temperature Below Threshold (Bank 1)
- P0430 Catalyst System Efficiency Below Threshold (Bank 2)
- P0431 Warm Up Catalyst Efficiency Below Threshold (Bank 2)
- P0432 Main Catalyst Efficiency Below Threshold (Bank 2)
- P0433 Heated Catalyst Efficiency Below Threshold (Bank 2)
- P0434 Heated Catalyst Temperature Below Threshold (Bank 2)

Figure 11-10 The CAT monitor can generate numerous DTCs.

ENABLING CRITERIA FOR THE CAT MONITOR

The enabling criteria for the CAT monitor are rather extensive but generally make sense because they focus on temperature, speed, and throttle. Let's look at a few examples of common vehicles. Figure 11-11 shows the enabling criteria for a 2007 Ford Escape with a 2.3 liter 4-cylinder engine.

Notice that the two "conditions" indicate a fuel tank between ½ and ¾ full, with ¾ full preferred. This is a frequent requirement of many monitors. At this fuel level there is minimal "fuel slosh" of the fuel in the tank. This allows the air/fuel ratio to be constant, a requirement for monitoring CAT efficiency. Condition #2 indicates that you are to operate the throttle smoothly during cruise conditions again to minimize fuel slosh. It is a good move to use cruise control if the vehicle is equipped with it, as this will minimize the action of the throttle. Flat-level road will also improve the ability to get this monitor to run.

2007 FORD ESCAPE: XLS 2.3L 153 HP L4 (Z) GAS-FI-N	
CATALYST MONITOR-ALL MODELS, ALL ENGINES EXC. DIESEL, HYBRID	
Conditions	1. Fuel tank should be 1/2–3/4 full; 3/4 full is preferable. 2. Operating the throttle smoothly during cruise or acceleration will minimize the time required for monitor completion and minimize "fuel slosh".
Step 1	Connect a scan tool to the data link connector. Turn the key ON with the engine OFF. Cycle the key OFF, then ON. Clear all DTCs and reset the PCM.
Step 2	Monitor the following PIDs: ECT, EVAPDC, FLI (if available) and TP MODE. Start the vehicle without returning to key OFF.
Step 3	Idle the engine for 15 seconds, then drive at 40 mph until the ECT is at least 170 degrees F.
Step 4	Drive in stop-and-go traffic. Include 5 different constant cruise speeds, ranging from 20–55 mph over a 10 minute period.
Step 5	Check monitor status on scan tool.
Step 6	If any non-continuous monitor has not completed, check for pending codes and make any necessary repairs. Re-run any incomplete monitors.

Delmar/Cengage Learning

Figure 11-11 Typical CAT enabling criteria.

The actual steps have some items that are very important for you to understand. If you are going to get this monitor to run, you will have to read between the lines. Notice step #3: "drive at 40 mph until ECT is at least 170 degrees F." Taken at face value, this means we would just go out and drive 40 mph. But reading between the lines helps us to realize that the ECT sensor must be accurate and the engine must have a functioning thermostat. In cold climates, a thermostat that is either leaking or stuck partially open will never allow the engine to reach 170 degrees. Once step #3 is achieved, step #4 brings up more and varied driving ("include 5 different cruise speeds ranging from 25–55 mph over a 10 minute period"). This may be the hardest component because we are to vary the speeds and yet achieve cruise conditions. Think about where you would drive around your repair shop that will allow speeds of 25 to 55 mph and be through by rush hour. You always have to determine where and when you can drive without traffic interfering with setting the monitors. Some technicians bring the vehicle home with them and take it out to try and run the monitors after dinner when the streets are less crowded. Make sure that you schedule time to run the monitors and get paid to set them.

Let's look at another example of a CAT monitor. Figure 11–12 shows the drive cycle for a 2005 Lexus GS300 with a 3.0 liter V-6.

This vehicle contains some slightly different enabling criteria and points out the obvious: you must have the correct enabling criteria for the year, make, and model of the vehicle or you may experience difficulties getting the monitor to run. Under the conditions notes it says that the MIL must be off. You will find this a common condition for non-continuous monitors. They will frequently not run if the MIL is on. This forces you to clear the codes prior to running the monitor, which will set all monitors to a not-run status. Be sure to only clear codes if the enabling criteria indicate that the monitor will not run if the MIL is on. It does not make sense to clear codes on every vehicle when you have only one or two monitors to run. Drive the enabling criteria for the not-run monitors only. Clearing the DTCs forces you to run all of the monitors, which could take substantially longer.

The second condition is engine temperature greater than 176 degrees F. Having a temperature condition is very common for CAT monitors. CATs are efficient only at operating temperatures, and if the engine is running too cold, it will have an impact on the monitor results. Think about this if you are in winter with a bad thermostat allowing the engine to run colder than specified. Leaking thermostats can and do result in false DTCs being set. Always make sure you match, or in this case exceed, the enabling criteria.

The last condition of greater than 14 degrees ambient temperature indicated on the IAT is easy to achieve but shows that the manufacturer wants the monitor to run only when both engine temperature and air temp are above a minimum temperature. On some vehicles an IAT reading substantially higher can shut off the monitor during the winter months. Some enabling criteria have printed values of greater than 45 degrees for the monitor to run. This effectively shuts off the monitor during winter operation in colder states.

The steps listed are very common and not difficult to achieve in most cases. Again you can see that Lexus wants the driver to operate the throttle smoothly and over a variety of speeds. Following this procedure should allow the monitor to run the prescribed two times, which will allow the readiness status screen to

CATALYST MONITOR (O2S TYPE) - GS300, GS430, LX470, ALL ENGINES

Notes	1. Monitor status will be rest to "Incomplete" if the ECU loses power, DTCs are cleared or conditions for enabling conditions have not been met. 2. If a drive cycle is interrupted, it can be resumed. 3. Avoid sudden speed changes.
Conditions	1. MIL must be OFF. 2. ECT > 176 degrees F. 3. IAT > 14 degrees F.
Step 1	Connect scan tool and check monitor status and preconditions.
Step 2	Note the IAT at engine start-up. The driving time must be adjusted depending on IAT at engine start-up.
Step 3	The readiness monitor can be completed at temperatures less than 14 degrees F, if the drive cycle is repeated a second time after cycling ignition OFF then ON again.
Step 4	Drive at 40–55 mph for 3 minutes. Drive smoothly and avoid sudden acceleration.
Step 5	If the IAT is less than 50 degrees F., drive at 40–55 mph for an additional 3 minutes.
Step 6	Drive the vehicle at 35–45 mph for approximately 7 minutes. Drive smoothly. Avoid sudden accelerations and sudden decelerations with throttle fully closed.
Step 7	Check the monitor status.
Step 8	If the readiness monitor does not switch to "Complete", the enabling criteria were probably not met. Turn the ignition OFF and repeat steps 4-6.
Step 9	The readiness status may not switch to "Complete" after the first drive cycle if a Pending DTC has been set. Once a second drive cycle is completed, a current DTC will be stored.
Step 10	If monitor does not switch to complete, turn the ignition OFF and repeat Steps 2–6.

Delmar/Cengage Learning

Figure 11-12 A CAT monitor that requires clearing DTCs (MIL must be turned off).

change to complete. Don't forget that the monitor status screen requires the monitor to run to completion twice to change. However, you will see in Chapter 15 that there is a scanner mode (mode 5/6) that allows a technician to look at the pieces of the monitor after the enabling criteria have been satisfied only once. This is very helpful when something prevents the monitor from running to completion.

Remember that the CAT monitor is probably the most important one to get to run because the CAT is one of the most critical components in the emission system. This is highlighted by the fact that some States require that the CAT monitor run if the vehicle has previously failed with a CAT code. The total number of monitors is still important, but the CAT has to have run or the vehicle is rejected.

CONCLUSION

In this chapter, we looked at the basics of the CAT monitor, emphasizing the functions of the oxygen sensors in determining the efficiency of the CAT. We discussed the typical signals coming off of the S1 sensors and compared them to the signals off of the S2 sensors. We analyzed how the ratio between the signals is used to determine how efficient the catalyst is at reducing the three pollutants: CO, HC, and NO_x. We emphasized that the CAT monitor is probably the most important system in keeping the vehicle within FTP specifications. We ended the chapter with a discussion of the enabling criteria necessary to get the monitor to run.

REVIEW QUESTIONS

1. Technician A states that a three-way converter can convert HC and CO. Technician B states that a three-way converter can convert NO_x. Who is correct?
   a. Technician A only
   b. Technician B only
   c. Both Technician A and B
   d. Neither Technician A nor B

2. Technician A states that HC and CO conversion is done at rich A/F ratios. Technician B states that NO_x conversion is done at lean A/F ratios. Who is correct?
   a. Technician A only
   b. Technician B only
   c. Both Technician A and B
   d. Neither Technician A nor B

3. Keeping the air/fuel ratio in the stoichiometric range is done by the

    a. mass air flow sensor

    b. S1 oxygen sensors

    c. S2 oxygen sensors

    d. engine coolant temperature sensor

4. The signal coming off of the S2 sensor shows lots of activity under cruise conditions. Technician A states that this indicates high CAT efficiency. Technician B states that this indicates low CAT efficiency. Who is correct?

    a. Technician A only

    b. Technician B only

    c. Both Technician A and B

    d. Neither Technician A nor B

5. The enabling criteria for the CAT monitor indicate an ECT temperature of at least 190 degrees. What is likely to happen if the engine temperature will not go above 160 degrees?

    a. The CAT monitor will run and generate a CAT DTC.

    b. The CAT monitor will run and not generate a CAT DTC.

    c. The CAT monitor will not run but will generate a CAT DTC.

    d. The CAT monitor will not run and will not generate any CAT DTCs.

6. Two technicians are discussing the enabling criteria for a CAT monitor and notice that cruising speeds in excess of 45 mph are required. Technician A states that this is to allow the CAT to reach operating temperature. Technician B states that cruising allows the signals from the O2Ss to stabilize. Who is correct?

    a. Technician A only

    b. Technician B only

    c. Both Technician A and B

    d. Neither Technician A nor B

7. A P0420 has been generated. A technician replaces the converter. How can he determine if the new catalyst is functioning at OEM standards?

    a. Clear the code to see if it comes back

    b. Use scan data off of various sensors to determine efficiency

    c. Have the vehicle emission tested

    d. Run the CAT monitor to completion

8. The enabling criteria have been met, and the CAT monitor will not run. Technician A states that some piece of data required by the enabling criteria might be missing. Technician B states that some piece of data required by the enabling criteria might not be to specification. Who is correct?

    a. Technician A only

    b. Technician B only

    c. Both Technician A and B

    d. Neither Technician A nor B

9. A vehicle generates an O2S DTC and a CAT efficiency DTC. Correct procedure involves which of the following?

    a. Diagnose and repair the O2S DTC.

    b. After the O2S repair, clear the DTCs and run the monitors.

    c. If the second running of the monitors generates a CAT DTC, diagnose and repair the CAT.

    d. All of the above

10. Technician A states that the CAT monitor's enabling criteria will usually run in city traffic as long as the engine coolant temperature is hot. Technician B states that getting the CAT monitor to run will involve all parts of the enabling criteria and specific driving loads and speeds. Who is correct?

    a. Technician A only

    b. Technician B only

    c. Both Technician A and B

    d. Neither Technician A nor B

EVAP MONITOR

OBJECTIVES

At the conclusion of this chapter you should able to: ■ Identify the enabling criteria for an EVAP system ■ Define an eight-hour cold soak and, using a scanner, determine why it will not occur ■ Recognize by looking at a diagram if an EVAP system is pressure or vacuum based ■ Understand the difference between a vacuum- and pressure-based EVAP monitor ■ Follow both purge and leak testing on a vacuum-based system ■ Recognize how a system detects a leak ■ Follow the wiring diagram for an EVAP system ■ Recognize and identify the conditions for setting a P0440 or a P0455 ■ Analyze a DSO pattern from a fuel tank pressure sensor ■ Explain the operation of the leak detection pump ■ Explain how a P0455 is set on a pressure-based system ■ Explain why having a DTC present might prevent a monitor from running

INTRODUCTION

Another non-continuous monitor is the EVAP monitor. Remember that non-continuous means that a specific set of conditions need to be present for the monitor to run. These conditions are the enabling criteria, and you must have access to them or you will likely not get the monitor to run. The EVAP system is responsible for eliminating the sources of unburned fuel being vented. Vented hydrocarbon is a large source of pollution. It is not a difficult system to design and build. However, the monitoring of its function is challenging. It is very common to see that the EVAP monitor has not run to completion during "normal" driving.

WHAT IS THE EVAP MONITOR?

The evaporative emissions monitor or EVAP is one of the most difficult monitors to get to run. Typically the enabling criteria involve conditions that may be difficult or even impossible for a consumer to get to run. Figure 12–1 shows the enabling criteria for a 1997 Jeep Grand Cherokee.

Notice that the first condition is "Cold soak for 8 hours without running engine," while the second is "Ambient temperature 40–90 degrees F; ECT within 10 degrees of ambient temperature." We have

discussed reading between the lines of the enabling criteria, and this is a great example. What does it mean to "cold soak" a vehicle? A cold soak has been achieved when the engine is completely cooled down to ambient or outside temperature. The reason Jeep says eight hours is because under normal conditions, that will be sufficient to completely cool down an engine. The eight-hour requirement is not cast in stone. What is cast in stone, however, is "ECT within 10 degrees of ambient temperature." Again reading between the lines, this means that the ECT temperature must be within 10 degrees of IAT. Intake air temperature will be ambient air temperature, and when the engine is completely cooled down, there will be a temperature match between IAT and ECT. Note that the PCM will not run the EVAP monitor if this temperature match has not occurred and ambient temperature is not between 40–90 degrees F. In cold winters and hot summers, the weather might cause the PCM to suspend the EVAP monitor. Ambient temperature requirements are not usually an issue except in seasonal extremes. However, the 10 degree ECT to IAT is a frequent problem. Figure 12–2 shows a scanner with both IAT and ECT PIDs listed.

This vehicle will not run the EVAP monitor or the CAT monitor and thus will not pass the required emission test. A further test of the actual engine

Notes	1. Disconnecting the battery will erase DTCs. 2. Once the monitor process has begun, do not turn the ignition key OFF. 3. Run monitors in the order of EVAP, Catalyst, EGR, O2 Sensor, Purge and O2 Sensor Heater.
Conditions	1. Cold soak for 8 hours without running engine. 2. Ambient temperature 40–90 degrees F.; ECT within 10 degrees F. of ambient temperature. 3. Fuel level 15–85% full.
Step 1	Chrysler Corp. recommends using the DRB III to run monitors on these vehicles. It is not necessary to start the engine to run this monitor. The EVAP monitor may also run while running the Catalyst monitor.
Step 2	Plug the DRB III or scan tool into the DLC.
Step 3	Let the vehicle sit overnight without starting the engine.
Step 4	Turn the ignition ON, but do not start the engine. The MIL should illuminate. If not, repair the MIL bulb.
Step 5	If using a DRB III, select #1 DRB III standalone.
Step 6	Select #1 1998–2002 diagnostics.
Step 7	Select #1 Engine.
Step 8	Select #2 DTCs and Related Functions.
Step 9	Select #1 and read DTCs. Monitors may not run and readiness will not be updated if DTCs are present. Repair and DTCs. Clear DTCs. OBD monitors will need to be run and completed to update readiness status.
Step 10	Return to Engine Select Function menu and select #9, OBDII monitors.
Step 11	Select #3 CARB readiness status.
Step 12	If the CARB readiness status reads YES, the vehicle is ready for an I/M test. It the CARB readiness status is No, run any monitors that do not read YES in the CARB readiness status. Continuous monitors are not included on CARB readiness status.
Step 13	Specific criteria need to be met for each monitor. Select EVAP LDP MON PRE-TEST from the DRB III, OBDII monitors menu.
Step 14	If using a generic scan tool, check the status of the EVAP monitor. If it is not ready, check for DTCs or that the enabling criteria have been met and repeat steps 1–3.

Delmar/Cengage Learning

Figure 12-1 The enabling criteria are sometimes difficult for the consumer.

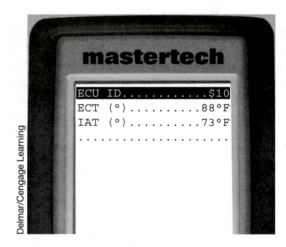

Delmar/Cengage Learning

Figure 12-2 If the difference on a cold start is more than 10 degrees F, the EVAP monitor will not run.

Delmar/Cengage Learning

Figure 12-3 After installation of a new ECT, the EVAP monitor ran to completion.

temperature indicated that the ECT was off by +9 degrees, always reading 9 degrees higher than actual engine temperature. At the same time, IAT was off by –6 degrees. This meant that under normal conditions with engine temperature and air temperature being the same, the PCM would see a 15 degree (–6 to +9)

difference and not run the EVAP monitor. In theory both sensors should be replaced; however, the shop decided to replace them one at a time. After installing a new ECT (Figure 12–3), the scanner showed the difference in ECT and IAT was now less than 10 degrees at start-up and the monitor ran. Even though the temperatures are not exact, as they should be, the difference of only 3 degrees allowed the monitor to run. Always pull up the enabling criteria when you are faced with a monitor that will not run, and read between the lines for the hidden information.

VACUUM-STYLE EVAP SYSTEMS

EVAP systems come in two varieties: pressure based and the more popular vacuum based. First, let's look at vacuum-based systems. As we examine the components and their function, keep in mind that the basic purpose of the EVAP system monitoring is to insure that no leaks are present that would allow unburned hydrocarbons to leak into the atmosphere. A loose of missing fuel cap is the best example of a large leak. Fuel evaporates and hydrocarbons rise because the system is not sealed. EVAP system testing is done to two standards, large and small. Large leak testing is designed to find leaks that are larger than .040". Small leak testing is designed to find leaks down to .020".

With vacuum-based EVAP, the PCM pulls a vacuum on the fuel system components, seals the system off, and makes sure that it holds a vacuum for a predetermined period of time. If it is capable of holding a vacuum, then it passes. If it cannot hold a vacuum, it fails and a DTC is set and the MIL turned on. Figure 12–4 shows a simplified EVAP system.

This simplified diagram shows the three important components: two valves, the vent solenoid and the purge solenoid; and one sensor, the fuel tank pressure sensor. While the fuel tank pressure sensor must be in the fuel tank, the other two valves may be placed just about anywhere on the vehicle with hoses connecting them to the system. When the EVAP system is tested, a sequence of events takes place that is PCM directed and involves a purge of the system and then a vacuum check. Following along with the diagram, let's look at purge first. With the vent solenoid open, as it is under normal no-power conditions, vapors from the gas tank are allowed to flow into the evaporative system canister. This is a container filled with charcoal that is connected to the area above the fuel in the tank. As the fuel expands or evaporates, the canister "catches" the vapors and holds them until purge occurs. Under specific conditions, the purge solenoid, which is normally closed, will be grounded by the PCM and open. This will allow engine vacuum to flow and pull vapors from the canister into the intake manifold, where they will be burned. The system is not under vacuum because the purge solenoid is open to the atmosphere. Once the canister has been purged of fuel vapors, the purge solenoid will be commanded closed by the PCM. The S1 oxygen sensors will indicate to the PCM that the purge has occurred. Remember that the S1 sensors are

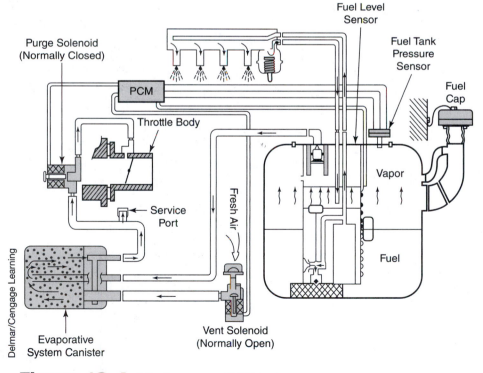

Figure 12-4 A basic vacuum EVAP system.

the ones in the exhaust manifold and closest to the cylinders. If the canister has significant fuel vapors stored and the purge opens, the S1 sensor(s) will swing rich momentarily. If the canister is not full of fuel vapors, the S1 sensors will swing lean. In either case, the change in the O2S signal is what indicates that the purge occurred. If the purge did not occur, the PCM will set a DTC.

The wiring diagram for a common GM EVAP system shows the relationship of the valve and its electrical connections (Figure 12–5).

The PCM is at the bottom of the diagram, with the two switch circuits for the purge solenoid and the vent solenoid. Notice that the PCM will apply a ground to the circuit to close the vent or open the purge. The vent valve is normally open with no ground applied, while the purge solenoid is opposite. It is normally closed and opened by the PCM by applying a ground. Both valves have power applied through two ignition-switched fuses.

When purge occurs, the signal from the front (S1) O2Ss indicates to the PCM if it occurred. If the PCM detects no purge, then a P0441 DTC should be set and the MIL turned on (Figure 12–6).

This P0441 is self-explanatory. Purge did not occur at a correct flow rate. You should be able to

access freeze frame and see the conditions that were present at the time the DTC set. Purge and leak testing are done separately. On some vehicles purge occurs first, and on others leak testing is done first. Either way it takes a pass on both for the system to be functional. Note that leak testing down to the .000 level is only done on "enhanced" systems. You can usually determine that the system is enhanced by the green cap under the hood, as Figure 12–7 shows. This cap allows factory testing equipment to be connected to the system.

Let's look at leak testing on the same vehicle. Remember that this is a vacuum-based system, which means that a vacuum will be pulled on the system and it has to hold. A loss in vacuum indicates a leak. There is a sequence of events that will take place during the running of the monitor. First the vent solenoid will be commanded to close. Figure 12–8 shows a vent solenoid in the rear wheel well close to the fuel tank.

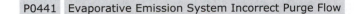

P0441 | Evaporative Emission System Incorrect Purge Flow

Figure 12-6 A DTC may indicate the result of a problem. Delmar/Cengage Learning

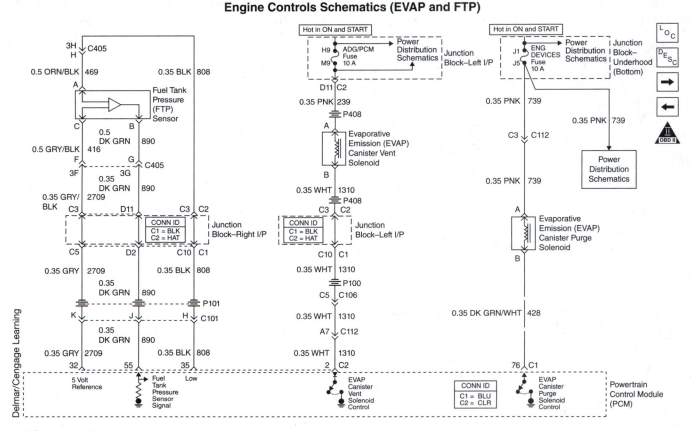

Figure 12-5 Wiring diagram for a GM EVAP system.

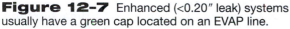
Green cap

Figure 12-7 Enhanced (<0.20″ leak) systems usually have a green cap located on an EVAP line.

Figure 12-8 This vent solenoid is located in a rear wheel well.

Notice the electrical connection and the one line. Under the cap shown in the middle of the picture is the actual vent opening. Once this vent is closed, the system is sealed. Next, the purge must open. The purge solenoid is usually located close to the throttle body, as Figure 12-9 shows.

There are two lines on this solenoid: one to the throttle body below the throttle plates (manifold vacuum) and one to the canister. With the engine running, the vent closed, and the purge open, manifold vacuum will draw down the entire system,

Figure 12-9 The purge solenoid is located under the hood near the throttle body.

including the fuel tank. Once the fuel tank pressure sensor indicates a strong vacuum, the purge will close, sealing off the system. The PCM now looks at the level of vacuum and should see it remain steady for a period of time, indicating no leaks. If a leak occurs, the vacuum will not remain steady. In operation, the fuel tank pressure sensor is similar to a common MAP sensor generating a voltage based on a differential between the inside tank pressure and atmospheric pressure. Figure 12-10 shows this relationship.

As the tank is placed under vacuum, the voltage rises and should hold once the purge is closed. Again, the rate of vacuum decay or loss is looked at over a period of time to determine the size of the leak. A solid system with no leaks will be seen as a sensor

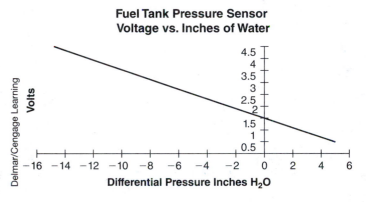

Figure 12-10 The fuel tank pressure sensor generates a voltage signal.

voltage that does not drop. There will be a PID for the fuel tank pressure sensor visible on a scanner like the one in Figure 12–11.

A DSO can also be used to view the voltage of the sensor over a period of time. In Figure 12–12, the vent valve has been closed and the purge opened. The maximum vacuum achieved a voltage of slightly over 4 volts. In the middle of the trace,

Figure 12-11 A scanner can show both the fuel tank sensor and the voltage.

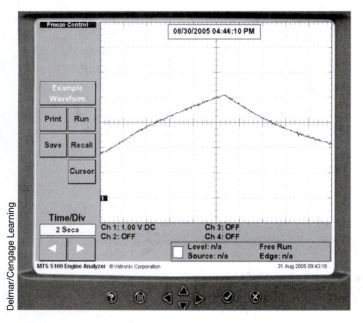

Figure 12-12 The DSO shows a significant pressure loss, indicating a leak.

the purge was closed and the voltage dropped off rapidly, indicating a major leak. If no leak was present, the trace would have been flat after the purge was closed.

PRESSURE-BASED SYSTEMS

A few vehicles have a pressure-based EVAP monitor system. The theory is basically the same with the exception that this system will check for leaks by pressurizing and watching for pressure loss. Remember that the vacuum-based system pulls a vacuum on the system and watches for the ability of the system to hold a vacuum. A loss in vacuum means that the system has a leak. In the pressure-based system, the system will be pressurized and a loss in pressure over a period of time will indicate a leak. Leaks are found through the loss of pressure.

The heart of the system is the leak detection pump (LDP) shown in Figure 12–13.

Notice that the canister vent valve is eliminated along with the fuel tank pressure sensor. The system still has the purge solenoid, which functions the same as in the vacuum-based system. The main difference is in the use of a leak detection pump, which we will examine in detail later. First, let's look at the purge function. As the fuel in the tank vaporizes, a line will carry the vapor directly to the canister. Under specific conditions, determined by the PCM, the purge solenoid will be energized and open. When it is open, vacuum from the intake manifold will flow through the valve to the canister, purging it of vapors. As with the vacuum-based system, the change in the signal from the O2S(s) will determine if the purge was functional. If the canister has significant fuel vapor, the O2S signal will momentarily indicate a rich condition. If the canister is void of fuel vapors, the signal from the O2S will momentarily indicate a lean condition. It is the swing from normal to either rich or lean that will indicate that purge has occurred. This function is part of the EVAP monitor, and requires that the O2S monitor has run successfully and did not generate any DTCs. Frequently, manufacturers will indicate that the EVAP monitor will not run if DTC are present and the MIL is on (Figure 12–14).

Figure 12–14 shows step 1 requiring the technician to clear the codes. This indicates that the manufacturers, Ford in this case, will not run the monitors if the MIL is on and DTCs are present. We stressed the importance of the O2Ss in previous chapters. EVAP is another system that will frequently utilize the signals from the front (S1) O2Ss and will not run if there are DTCs.

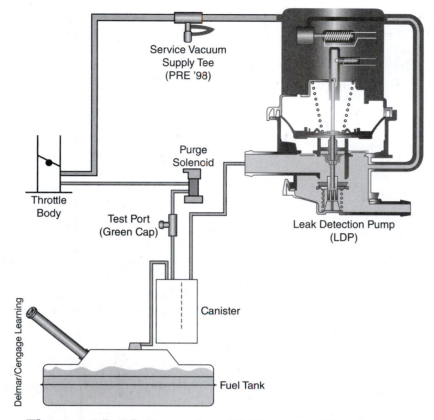

Figure 12-13 Pressure-based EVAP systems utilize a leak detection pump.

Step 1	Connect a scan tool to the data link connector. Turn the key ON with the engine OFF. Cycle the key OFF, then ON. Clear all DTCs and reset the PCM.
Step 2	Monitor the following PIDs: ECT, EVAPDC, FLI (if available) and TP MODE. Start the vehicle without returning to key OFF.
Step 3	Idle the engine for 15 seconds, then drive at 40 mph until the ECT is at least 170 degrees F.
Step 4	If the IAT is between 40–100 degrees F., go to step 5. If not, complete each of the following steps. Note that Step 14 is required to bypass the EVAP monitor and clear the P1000 code.
Step 5	See HEGO Monitor Drive Cycle, Step 4.
Step 6	See EVAP Monitor Drive Cycle, Steps 4–5.
Step 7	See Catalyst Monitor Drive Cycle, Step 4.
Step 8	See EGR Monitor Drive Cycle, Steps 4–5.

Figure 12-14 Step 1 of the enabling criteria indicates to "clear all DTCs."

Leak detection can occur either prior to purge or after, based on the sequence of the monitor steps. Refer to Figure 12–15 for the following discussion.

There are two sets of electrical connections on the LDP. The top ones are for a solenoid valve, which will allow engine vacuum to the pump. The second is a micro-switch that will indicate the position of the pumping diaphragm. Both of these electrical connections come from the PCM. When the PCM sees that the correct enabling criteria are present, it will close the purge if it is open, and open up the solenoid vacuum valve. Vacuum will pull up the diaphragm against spring tension. When the diaphragm is all the way in the upward position, the micro-switch will close. This signals the PCM to close the solenoid vacuum valve. With no vacuum holding the diaphragm up, the spring will push down, pressurizing the lower chamber, which is connected to the canister and fuel tank. This pumping process will continue over and over until a balance between the spring tension and system pressure occurs. Because the system is pressurized, the pumping action stops. This occurs at a pressure of 7.5 inches water. Note: This is not 7.5 PSI; 7.5 inches

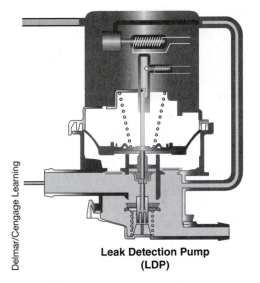

Leak Detection Pump (LDP)

Figure 12-15 The heart of a pressure-based system is the leak detection pump.

water actually equals only .27 PSI. Once the pumping stops, the PCM determines how long it took to achieve the 7.5 in H_2O. The PCM knows the amount of fuel in the tank and therefore the volume of space above the fuel, so it calculates how long the LDP should pump to pressurize this volume. Additionally, if the pumping never stops, it will set a large leak DTC (P0455).

Note the large number of possible DTCs generated off of the EVAP monitor, shown in Figure 12–16. This accounts for the difficulty in getting the monitor to run frequently. The more "pieces" of the system that are being tested, the more complicated the enabling criteria become and the more difficult it is to get the monitor to run.

Let's get back to our LDP. Once the system is pressurized, the pump stops. If there is a leak, the pressure will be reduced below the 7.5 in H_2O pressure of the spring. When this occurs, the pump will begin

P043E	Evaporative Emission System Leak Detection Reference Orifice Low Flow
P043F	Evaporative Emission System Leak Detection Reference Orifice High Flow
P0440	Evaporative Emission System
P0441	Evaporative Emission System Incorrect Purge Flow
P0442	Evaporative Emission System Leak Detected (small leak)
P0443	Evaporative Emission System Purge Control Valve Circuit
P0444	Evaporative Emission System Purge Control Valve Circuit Open
P0445	Evaporative Emission System Purge Control Valve Circuit Shorted
P0446	Evaporative Emission System Vent Control Circuit
P0447	Evaporative Emission System Vent Control Circuit Open
P0448	Evaporative Emission System Vent Control Circuit Shorted
P0449	Evaporative Emission System Vent Control Circuit Intermittent
P0450	Evaporative Emission System Pressure Sensor/Switch
P0451	Evaporative Emission System Pressure Sensor/Switch Range/Performance
P0452	Evaporative Emission System Pressure Sensor/Switch Low
P0453	Evaporative Emission System Pressure Sensor/Switch High
P0454	Evaporative Emission System Pressure Sensor/Switch Intermittent
P0455	Evaporative Emission System Leak Detected (gross leak/no flow)
P0456	Evaporative Emission System Leak Detected (very small leak)
P0457	Evaporative Emission System Leak Detected (fuel cap loose/off)
P0458	Evaporative Emission System Purge Control Valve Circuit Low
P0459	Evaporative Emission System Purge Control Valve Circuit High

Figure 12-16 EVAP DTCs can be very specialized.

working. The PCM can look at how long the pump was off to determine the size of the leak and set the corresponding DTC. Once leak detection is over, the purge will be opened to release the pressure.

CONCLUSION

You have learned from the discussion that leak detection can be pressure based or vacuum based, with each system having the ability to determine the size of the leak. One of the most common DTC found in the emission testing lanes is "P0455 – Large Leak Detected." Frequently, it is nothing more than a loose gas cap, but it can be any part of the EVAP system. The other thing to realize about EVAP is that both continuous and non-continuous monitors are at work. The continuous, comprehensive component monitor checks the power and ground circuits and the electrical solenoids like the purge valve and the vent valve, while the non-continuous EVAP monitor checks the actual flow through purge and does leak detection. When a PCM detects a problem and sets a DTC, the technician must look at the DTC and determine what monitor set it. This information will be helpful especially in baseline and repair verification.

REVIEW QUESTIONS

1. Two technicians are discussing the enabling criteria for an EVAP monitor. Technician A states that the criteria might contain an eight-hour cold soak. Technician B states that the monitor will use the input from some sensor to indicate a leak. Who is correct?
   a. Technician A only
   b. Technician B only
   c. Both Technician A and B
   d. Neither Technician A nor B

2. The enabling criteria call for an eight-hour cold soak. Scan data will show this has been accomplished by
   a. a cold soak PID
   b. ECT and IAT being close to equal
   c. complete engine cooldown in over 60 degrees outside temperature
   d. the presence of a cold soak DTC

3. Technician A states that a system might use engine vacuum to draw down an EVAP system and watch how long it can hold vacuum. Technician B states that a system might use engine compression pressure to pressurize the EVAP system and watch for leaks. Who is correct?
   a. Technician A only
   b. Technician B only
   c. Both Technician A and B
   d. Neither Technician A nor B

4. Purge is accomplished on a vacuum-based system by
   a. opening the purge valve and closing the vent valve
   b. closing the purge valve and closing the vent valve
   c. closing the purge valve and opening the vent valve
   d. opening the purge valve and opening the vent valve

5. Most systems know that purge occurred by looking at the
   a. O2S signal
   b. ECT signal
   c. purge valve ground
   d. tank pressure sensor

6. A vehicle comes in with a continuously running leak detection pump. Technician A states that this might indicate an open ground circuit from the pump. Technician B states that this might indicate a short to ground on the ground side of the pump. Who is correct?
   a. Technician A only
   b. Technician B only
   c. Both Technician A and B
   d. Neither Technician A nor B

7. A DSO on the fuel tank pressure sensor shows no change in voltage with the vent valve closed and the purge valve open (engine running). Technician A states that this might indicate a large leak. Technician B states that it might indicate the purge valve did not open. Who is correct?
   a. Technician A only
   b. Technician B only
   c. Both Technician A and B
   d. Neither Technician A nor B

8. After a vehicle has been off for 12 hours, the ECT temperature is 42 degrees F. The IAT temperature is 30 degrees F. Technician A states that this might prevent the EVAP monitor from running.

Technician B states that either the IAT or the ECT is not correctly indicating temperature. Who is correct?

a. Technician A only

b. Technician B only

c. Both Technician A and B

d. Neither Technician A nor B

9. During the running of the EVAP monitor, the leak detection pump runs and does not turn off. This most likely indicates

a. a faulty leak detection pump

b. a PCM failure

c. a large leak within the EVAP system

d. a closed purge valve

10. A P0441 is present. Technician A states that the purge valve might not be functional. Technician B states that this indicates that the EVAP monitor must not have run to completion. Who is correct?

a. Technician A only

b. Technician B only

c. Both Technician A and B

d. Neither Technician A nor B

EGR MONITOR

CHAPTER 13

OBJECTIVES

At the conclusion of this chapter you should be able to: ■ Understand how bleeding exhaust gas into the intake manifold helps to reduce the production of NO_x ■ Recognize, by looking at a diagram, how the system will control and measure the amount of exhaust flowing ■ Analyze the DTC present and determine where to begin the diagnostics ■ Follow a wiring diagram for an EGR system ■ Determine how the EGR system will know if a flow problem is present ■ Recognize the correlation between EGR flow and NO_x ■ Diagnose the basics of the Ford DPFE ■ Run the enabling criteria to completion for an EGR system ■ Diagnose or analyze the use of a MAP to determine flow ■ Diagnose or analyze the use of the O2S to determine flow ■ Diagnose or analyze the use of an exhaust backpressure transducer

INTRODUCTION

Another very important monitor within the OBD II system is the exhaust gas recirculation (EGR) monitor. The EGR system, although not used on all vehicles, is primarily responsible for maintaining combustion temperatures below 2500 degrees F. If the combustion process were to develop temperatures in excess of 2500 degrees, there would be an increase in the NO_x levels. This increase would put a strain on the CAT. The EGR system bleeds some exhaust gas into the intake manifold to dilute the air/fuel ratio. The exhaust gas cannot burn again and will not increase the combustion temperature.

HOW THE EGR SYSTEM WORKS

The first thing we have to do prior to understanding how the monitor will function is to look at how the typical system actually works. Keep in mind that there are many different styles of EGR systems, but they can be divided into two major styles, vacuum operated or electrically operated. Let's look at the vacuum system first. Figure 13–1 shows a typical system.

The EGR valve labeled #1 is the heart of the system. It has a vacuum line running to the top of the valve (#7) and two large tubes that will carry the exhaust gas from the exhaust manifold (#3) to the intake manifold (#2). When exhaust gas is required, vacuum

EGR System (24 V)

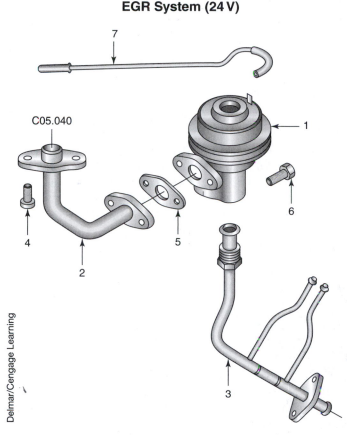

Delmar/Cengage Learning

Figure 13-1 A typical non-monitored EGR system.

will be ported over to the valve, allowing it to open. When the valve is open, the combination of vacuum in the intake manifold and pressure in the exhaust manifold will cause exhaust gas to flow. The exhaust gas will mix with the air/fuel mixture and dilute it. The system is simple and yet very prone to problems. The passages clog up, and the vacuum regulation allows either too much or too little exhaust, causing drivability and/or emissions problems. The EGR system has been around for many years but cannot be called a favorite system of technicians. This is primarily due to the drivability problems that have plagued the system. Once the EGR system was computerized, many of those problems were cured. Adding monitoring of the system in 1996 was another improvement, because the OBD II system could now catch some problems prior to their becoming an issue. The system is monitored to prevent emission problems. Remember if the system does not work, NO_x will increase; and if the system flows too much exhaust, misfire will occur, which increases the level of HC going into the CAT. The amount of EGR must be correct. How each manufacturer monitors the amount of exhaust will vary and will result in a DTC when not correct.

There are many DTCs that can be set indicating EGR system problems. Some of the DTCs are flow oriented like the P0401, and others are component oriented like the P0405. Figure 13–2 shows most of the possible EGR system DTCs.

Although the list is extremely long, you need to recognize that the most common EGR DTC is a P0401. The top 20 DTCs are listed in Figure 13–3.

Notice that the second most common DTC was "P0401—Exhaust Gas Recirculation Flow Insufficient Detected." When you display the top 20 in pie chart format, it shows the high frequency (12%) that P0401 has (Figure 13–4). Most years in most States will have a P0401 in the top 5 common DTC list.

As we go through the basics of the EGR monitor, always keep in mind that the function of the EGR system is to bleed some exhaust gas into the intake mixture to keep the burn temperature below the 2500 degree threshold that allows NO_x to be produced.

FORD EGR SYSTEM AND MONITOR

The Ford system is a good one to begin the specific systems with. It is the only system that actually measures the amount of EGR gas that is flowing through the use of a DPFE (differential pressure feedback EGR), shown in Figure 13–5.

Follow along with the illustration. A pulse width modulated signal is sent to the EGR vacuum regulator solenoid, which will control the source vacuum to the EGR valve. The EGR valve is a simple vacuum valve that will control the flow of exhaust gas from the exhaust manifold to the intake manifold. When vacuum is applied to the top of the valve, it will open and exhaust will flow. When the PCM shuts off the EGR vacuum regulator solenoid, a spring inside the EGR valve will shut off the flow of exhaust. This system, so far, is not complicated and very common in the early '80s. The difference comes in the use of a sensor to determine the amount of EGR flow by looking at the pressure drop across a calibrated orifice. The orifice creates a pressure difference, which is measured by the DPFE and indicates the quantity of exhaust gas actually flowing. The voltage signal is sent to the PCM and used by it to modify the pulse width to the EGR vacuum regulator solenoid. There is a PID for the DPFE signal that can be observed while driving the vehicle. This gives the technician the ability to determine if there is any EGR flow. By observing the PID for the EGR vacuum regulator solenoid, a comparison can be made between the solenoid signal and the DPFE voltage to determine if the two seem to match. When the passages become partially blocked, EGR is reduced and the DPFE voltage drops. Running the monitor is not too difficult, as seen in Figure 13–6.

Notice the usual fuel level requirement and smooth operation of the throttle listed under conditions. There is no eight-hour soak, or strange conditions that will be difficult to achieve. The monitor is pretty straightforward for this 2002 Crown Victoria. When you look at the steps, you will realize that the only requirement is a warm engine (>170 degrees) and some acceleration to 45 mph at ½ to ¾ throttle. By the time the vehicle has been accelerated three times, the monitor should run. It is very uncommon to find the EGR monitor not run under normal driving conditions because it has a very simple set of enabling criteria. However, as it runs, it is frequently the source of DTCs that will need to be repaired prior to running the monitor again.

The key to the Ford system is the DPFE and how it functions. There have been numerous design changes in the DPFE, so always check the applicable TSBs to determine the correct unit. There are few manufacturers that will measure the amount of exhaust gas flowing directly. Most will look at the O2S or fuel trim or some other indicator of exhaust flow and indirectly calculate the quantity.

P0400	Exhaust Gas Recirculation "A" Flow
P0401	Exhaust Gas Recirculation "A" Flow Insufficient Detected
P0402	Exhaust Gas Recirculation "A" Flow Excessive Detected
P0403	Exhaust Gas Recirculation "A" Control Circuit
P0404	Exhaust Gas Recirculation "A" Control Circuit Range/Performance
P0405	Exhaust Gas Recirculation Sensor "A" Circuit Low
P0406	Exhaust Gas Recirculation Sensor "A" Circuit High
P0407	Exhaust Gas Recirculation Sensor "B" Circuit Low
P0408	Exhaust Gas Recirculation Sensor "B" Circuit High
P0409	Exhaust Gas Recirculation Sensor "A" Circuit
P040A	Exhaust Gas Recirculation Temperature Sensor "A" Circuit
P040B	Exhaust Gas Recirculation Temperature Sensor "A" Circuit Range/Performance
P040C	Exhaust Gas Recirculation Temperature Sensor "A" Circuit Low
P040D	Exhaust Gas Recirculation Temperature Sensor "A" Circuit High
P040E	Exhaust Gas Recirculation Temperature Sensor "A" Circuit Intermittent/Erratic
P040F	Exhaust Gas Recirculation Temperature Sensor "A" / "B" Correlation
P041A	Exhaust Gas Recirculation Temperature Sensor "B" Circuit
P041B	Exhaust Gas Recirculation Temperature Sensor "B" Circuit Range/Performance
P041C	Exhaust Gas Recirculation Temperature Sensor "B" Circuit Low
P041D	Exhaust Gas Recirculation Temperature Sensor "B" Circuit High
P041E	Exhaust Gas Recirculation Temperature Sensor "B" Circuit Intermittent/Erratic
P042E	Exhaust Gas Recirculation "A" Control Stuck Open
P042F	Exhaust Gas Recirculation "A" Control Stuck Closed
P044A	Exhaust Gas Recirculation Sensor "C" Circuit
P044B	Exhaust Gas Recirculation Sensor "C" Range/Performance
P044C	Exhaust Gas Recirculation Sensor "C" Circuit Low
P044D	Exhaust Gas Recirculation Sensor "C" Circuit High
P044E	Exhaust Gas Recirculation Sensor "C" Circuit Intermittent/Erratic

Figure 13-2 The list of EGR DTCs is extensive and very specific. *(continued)*

Delmar/Cengage Learning

P045A	Exhaust Gas Recirculation "B" Control Circuit
P045B	Exhaust Gas Recirculation "B" Control Circuit Range/Performance
P045C	Exhaust Gas Recirculation "B" Control Circuit Low
P045D	Exhaust Gas Recirculation "B" Control Circuit High
P045E	Exhaust Gas Recirculation "B" Control Stuck Open
P045F	Exhaust Gas Recirculation "B" Control Stuck Closed
P046C	Exhaust Gas Recirculation Sensor "A" Range/Performance
P046D	Exhaust Gas Recirculation Sensor "A" Intermittent/Erratic
P046E	Exhaust Gas Recirculation Sensor "B" Range/Performance
P046F	Exhaust Gas Recirculation Sensor "B" Intermittent/Erratic
P0470	Exhaust Pressure Sensor "A" Circuit
P0471	Exhaust Pressure Sensor "A" Circuit Range/Performance
P0472	Exhaust Pressure Sensor "A" Circuit Low
P0473	Exhaust Pressure Sensor "A" Circuit High
P0474	Exhaust Pressure Sensor "A" Circuit Intermittent/Erratic
P0475	Exhaust Pressure Control Valve "A"
P0476	Exhaust Pressure Control Valve "A" Range/Performance
P0477	Exhaust Pressure Control Valve "A" Low
P0478	Exhaust Pressure Control Valve "A" High
P0479	Exhaust Pressure Control Valve "A" Intermittent
P047A	Exhaust Pressure Sensor "B" Circuit
P047B	Exhaust Pressure Sensor "B" Circuit Range/Performance
P047C	Exhaust Pressure Sensor "B" Circuit Low
P047D	Exhaust Pressure Sensor "B" Circuit High
P047E	Exhaust Pressure Sensor "B" Circuit Intermittent/Erratic
P047F	Exhaust Pressure Control Valve "A" Stuck Open
P0486	Exhaust Gas Recirculation Sensor "B" Circuit
P0487	Exhaust Gas Recirculation Throttle Control Circuit "A" / Open
P0488	Exhaust Gas Recirculation Throttle Control Circuit "A" Range/Performance

Delmar/Cengage Learning

Figure 13-2 *(continued)*

P0489	Exhaust Gas Recirculation "A" Control Circuit Low
P048A	Exhaust Pressure Control Valve "A" Stuck Closed
P048B	Exhaust Pressure Control Valve "A" Position Sensor/Switch Circuit
P048C	Exhaust Pressure Control Valve "A" Position Sensor/Switch Circuit Range/Performance
P048D	Exhaust Pressure Control Valve "A" Position Sensor/Switch Circuit Low
P048E	Exhaust Pressure Control Valve "A" Position Sensor/Switch Circuit High
P048F	Exhaust Pressure Control Valve "A" Position Sensor/Switch Circuit Intermittent/Erratic
P0490	Exhaust Gas Recirculation "A" Control Circuit High
P049A	Exhaust Gas Recirculation "B" Flow
P049B	Exhaust Gas Recirculation "B" Flow Insufficient Detected
P049C	Exhaust Gas Recirculation "B" Flow Excessive Detected
P049D	Exhaust Gas Recirculation "A" Control Position Exceeded Learning Limit
P049E	Exhaust Gas Recirculation "B" Control Position Exceeded Learning Limit
P049F	Exhaust Pressure Control Valve B
P04A0	Exhaust Pressure Control Valve "B" Range/Performance
P04A1	Exhaust Pressure Control Valve "B" Low
P04A2	Exhaust Pressure Control Valve "B" High
P04A3	Exhaust Pressure Control Valve "B" Intermittent
P04A4	Exhaust Pressure Control Valve "B" Stuck Open
P04A5	Exhaust Pressure Control Valve "B" Stuck Closed
P04A6	Exhaust Pressure Control Valve "B" Position Sensor/Switch Circuit
P04A7	Exhaust Pressure Control Valve "B" Position Sensor/Switch Circuit Range/Performance
P04A8	Exhaust Pressure Control Valve "B" Position Sensor/Switch Circuit Low
P04A9	Exhaust Pressure Control Valve "B" Position Sensor/Switch Circuit High
P04AA	Exhaust Pressure Control Valve "B" Position Sensor/Switch Circuit Intermittent/Erratic

Figure 13-2 (continued)

Illinois Top 20 Codes 2007

Code	Type	Count	Description
P0420	NC	11614	Catalyst System Efficiency Below Threshold (Bank 1)
P0171	C	10434	System Too Lean (Bank 1)
P0401	NC	9382	Exhaust Gas Recirculation Flow Insufficient Detected
P0174	C	7349	System Too Lean (Bank 2)
P0440	NC/C	5525	Evaporative Emission Control System
P0300	C	5515	Random Misfire Detected
P0442	NC	5166	Evaporative Emission Control System Leak Detected (small leak)
P0141	NC	4390	Oxygen Sensor Heater Circuit (Bank 1, Sensor 2)
P0455	NC	4283	Evaporative Emission Control System Leak Detected (gross leak)
P0430	NC	3888	Catalyst System Efficiency Below Threshold (Bank 2)
P0133	NC	3352	Oxygen Sensor Circuit Slow Response (Bank 1, Sensor 1)
P0446	C	3281	Evaporative Emission Control System Vent Control Circuit
P0301	C	3006	Cylinder 1 Misfire Detected
P0304	C	2856	Cylinder 4 Misfire Detected
P0441	NC	2770	Evaporative Emission Control System Incorrect Purge Flow
P0135	NC	2720	Oxygen Sensor Heater Circuit (Bank 1, Sensor 1)
P0303	C	2602	Cylinder 3 Misfire Detected
P0302	C	2545	Cylinder 2 Misfire Detected
P0325	C	2279	Knock Sensor 1 Circuit (Bank 1 or Single Sensor)
P0134	NC	1788	Oxygen Sensor Heater Circuit Insufficient Activity (Bank 1, Sensor 1)

Delmar/Cengage Learning

Figure 13-3 The top 20 DTCs always contain a P0401, indicating EGR flow insufficient detected.

Illinois 2005 Top 20 Diagnostic Trouble Codes

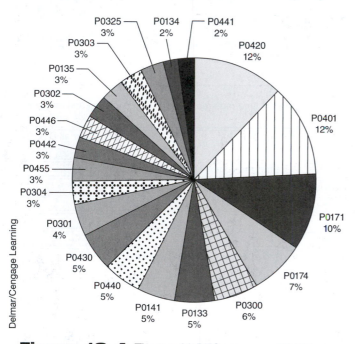

Figure 13-4 The top 20 DTCs shows a P0401 at 12% overall.

GM EGR SYSTEM AND MONITOR

The Ford EGR valve was vacuum operated and straightforward. Let's look at a fully electronic version of the EGR valve. We will use the GM version as an example. Figure 13–7 shows an EGR valve on a 6-cylinder engine.

The main difference in this valve is that it is fully electric. Notice that there is no vacuum line. The PCM can directly control the valve by pulse width modulating the signals to the internal solenoid. There is a position sensor mounted in the top of the valve.

Notice that the PCM is in control of the ground side of the EGR circuit. There is a 20 amp fuse powering the feed side of the EGR solenoid in Figure 13–8.

The solenoid is independent with a power feed and a ground circuit within the PCM and will be pulse width modulated to achieve the desired flow. Because this valve is completely electronic, you will find a scanner PID for position and usually one for flow in percent. The last piece of the puzzle is the position sensor, which is usually mounted in the top of

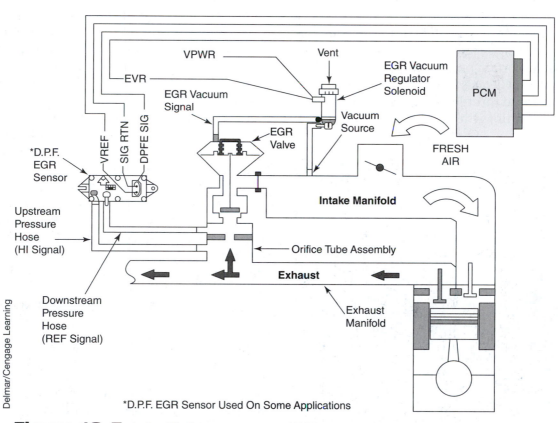

*D.P.F. EGR Sensor Used On Some Applications

Figure 13-5 A simplified vacuum-operated EGR system with electronic monitoring.

OBD II Drive Cycles		Ford
		Crown Victoria

All Monitors		**2002**
Conditions	1. Fuel tank should be 1/2–3/4 full; 3/4 full is preferable. 2. Operating the throttle smoothly during cruise or acceleration will minimize the time required for monitor completion and minimize "fuel slosh."	
Step 1	Connect a scan tool to the data link connector. Turn the key ON with the engine OFF. Cycle the key OFF, then ON. Clear all DTCs and reset the PCM.	
Step 2	Monitor the following PIDs: ECT, EVAPDC, FLI (if available) and TP MODE. Start the vehicle without returning to key OFF.	
Step 3	Idle the engine for 15 seconds, then drive at 40 mph until the ECT is at least 170 degrees F.	
Step 4	From a stop, accelerate to 45 mph at 1/2–3/4 throttle.	
Step 5	Repeat Step 4 three times.	
Step 6	Check monitor status on scan tool.	
Step 7	If any non-continuous monitor has not completed, check for pending codes and make any necessary repairs. Re-run any incomplete monitors.	

Figure 13-6 The enabling criteria for a Ford EGR system.

the valve. Functionally it is similar to a throttle position sensor, so a reading of .5 V would indicate that the valve is not open; 2.5 V would indicate 50% flow; and 4.5 V would indicate 100% flow (almost never achieved). GM is a good example of a digital valve with position sensing built in. The enabling criteria for a GM vehicle running the EGR monitor are not difficult to achieve, as Figure 13–9 shows.

The most important part of these enabling criteria is the first note: "EGR tests run during a gradual deceleration with closed throttle at speeds above 30 mph." Note that this is not when the EGR flow normally

Figure 13-7 An all-electronic EGR contains solenoids and position monitoring.

takes place; this is when the monitor runs. So several deceleration runs with closed throttle might be necessary. Remember also that, most likely, the monitor will have to run twice before the monitor status screen will change from not run to run. There are examples of the monitor running once, determining that there is a problem, not setting a DTC, and shutting down the second running of the monitor. This prevents the monitor status screen from switching over to a run status. As a technician, you will find examples where you will have to look at the design of the monitor to determine why it will not run to completion. We will look at Mode 5 and 6 in chapter 15 and see how they might be useful when the monitor runs once but not the required second time.

Remember the Ford system with its DPFE indicating EGR flow? Every monitor has to be able to detect at EGR flow in one form or another. GM does it by analyzing a MAP (manifold absolute pressure) even in systems that use a MAF (mass air flow) to indicate air flow. Prior to OBD II, many vehicles used a speed density system to determine the amount of air flowing into the engine. The MAP signal was used to determine the pressure in the manifold, and the signal off of the crank gave the PCM the speed reference it needed. The MAF meter replaced the MAP sensor on most applications. When you see a MAP on a GM vehicle, in addition to a MAF, it is there for the EGR flow requirement of the monitor. As the EGR valve functions, the pressure in the intake manifold will change and the MAP will indicate the change to the PCM. If for some reason the MAP on a GM is

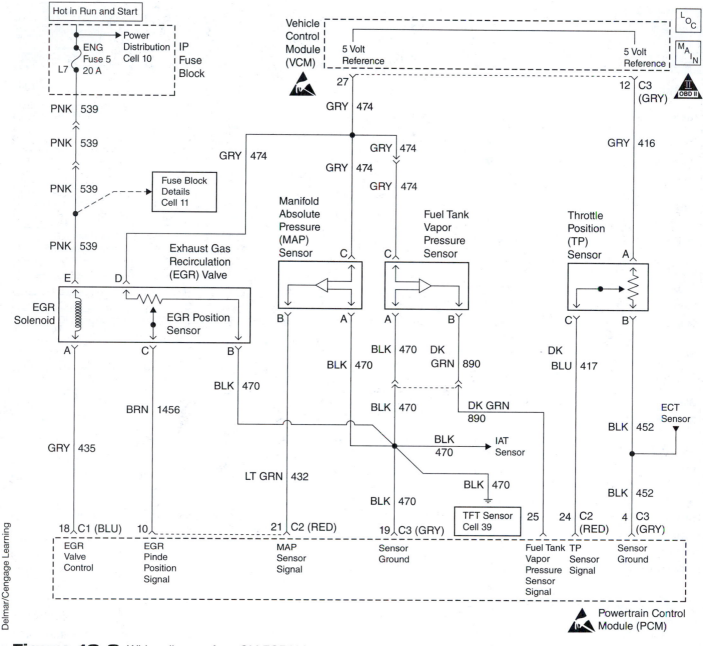

Figure 13-8 Wiring diagram for a GM EGR Valve.

not functional, the EGR monitor will shut down. Figure 13–10 shows a common GM MAP.

It will have three wires: one is 5-volt power, one is ground, and one is the signal back to the PCM. As the EGR is controlling the amount of exhaust flow desired, the MAP will act as a messenger, sending back the change in manifold vacuum. It is this change that will indicate the volume or amount of exhaust being mixed in the intake manifold. The different DTCs available will be able to identify the sections of the EGR system, so diagnosis will be directed toward the real problem. Remember that any DTC with the

word "circuit" in it indicates that the comprehensive component monitor has generated the DTC. This same monitor will be able to indicate MAP DTCs. However, it is the EGR monitor, which is part of the non-continuous monitor group, that will be capable of indicating flow difficulties. The very common "P0401—Exhaust Gas Recirculation Flow Insufficient" is set by running the non-continuous EGR monitor to completion twice. If the monitor will only run once, it might generate a pending code that might not be shown on some inexpensive generic-only scanners. Make sure your scanner can display both generic

OBD II Drive Cycles

Chevrolet
Impala

All Monitors		2005
Notes	1. EGR tests run during a gradual deceleration with closed throttle at speeds above 30 mph. Several deceleration cycles may be necessary to accumulate enough EGR flow samples. 2. Drive cycle may abort due loss of, changed, enabling conditions.	
Conditions	1. ECT > 167–258 degress F.; IAT 32–212 degrees F.; engine speed 1000–1400 rpm. 2. MAP 17–43 kPa and does not change >1.1 kPa; TP sensor <1%; 3. Ignition 1 voltage 11–18V; BARO > 74 kPa; EGR position <1%	
Step 1	Perform the I/M System Check. Failure to do so may result in difficulty in updating the status to YES. If the monitor does not reset after numerous attempts, and no DTC is set, it is probable that the enabling criteria have not been met.	
Step 2	Ensure all conditions have been met - traction control inactive, A/C Relay command and Current Gear parameters should not change, MAF should not change > 2g/s, IAC position should not change more than 5 counts and Decel Fuel Cutoff mode OFF.	
Step 3	Turn OFF all accessories.	
Step 4	Start the engine and idle for 2 minutes.	
Step 5	Accelerate at part throttle to at least 55 mph and maintain speed for 1 minute.	
Step 6	Decelerate to 30 mph (do not go below 30 mph) with the throttle closed, NO brake application, NO clutch actuation and NO manual downshift.	
Step 7	If the monitor is YES, go to Step 15. If not, go to Step 8.	
Step 8	Using a scan tool, access the DTC information. If there are any failed DTCs, diagnose and repair the DTCs. If no failed DTCs, go to next step.	
Step 9	Determine which DTCs are required to run in order to complete the test.	
Step 10	Using a scan tool, observe the Not Ran Since Code Cleared display. Determine which DTCs that are required to have run, and have not run, for a YES status to be displayed.	
Step 11	Enter the DTC in the specific DTC menu of the scan tool.	
Step 12	Be sure that all of the monitor Enabling Conditions have been met. Repeat the procedure until the scan tool indicated the diagnostic test has run.	
Step 13	Repeat Steps 11–12 for any DTCs that have not run.	
Step 14	Check the I/M System Readiness display. If the monitor status is YES, go to Step 15. If not perform diagnosis and repair.	
Step 15	Check for emission-related DTCs. If any are present, go back to Step 8. If no DTCs are present, EGR system status is OK.	

Delmar/Cengage Learning

Figure 13-9 The enabling criteria for a GM vehicle EGR system.

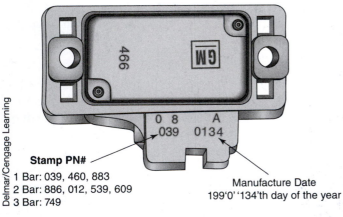

Delmar/Cengage Learning

Stamp PN#
1 Bar: 039, 460, 883
2 Bar: 886, 012, 539, 609
3 Bar: 749

Manufacture Date
199'0''134'th day of the year

Figure 13-10 Sometimes EGR flow is measured by using a MAP sensor.

and manufacturer's information. States with required emission testing rely on the generic data stream, so you must have access to it. However, the data shown on the manufacturer's side of the system is more extensive and frequently leads to faster diagnosis and repair. Once the repair is completed, switch over to generic to check on monitor status and make sure the vehicle is ready for the emission test. Remember, monitor status, MIL function, and P0 DTC "live" in the generic side of the system. You must have the capability of looking at generic data, especially when a passed emission test is the goal.

CHRYSLER VACUUM EGR SYSTEM AND MONITOR

Chrysler uses a few different types of EGR systems. The newest one is almost exactly the same as the GM. However, there are many vacuum systems around. Chrysler made the effort to totally eliminate the EGR system on cars by 2007/08. The vacuum system with a transducer was around for many years and was computerized for OBD II in 1996. Figure 13–11 shows the system.

Chrysler Backpressure EGR System

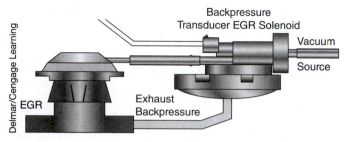

Figure 13-11 Backpressure tranducers are sometimes used to help control the EGR valve.

The control of any EGR valve remains the biggest issue, and Chrysler chose to use exhaust backpressure to indicate how much exhaust could be dumped into the manifold. The line from under the EGR valve indicates the level of backpressure in the exhaust system. The transducer uses this pressure to control the opening and closing of the valve. The two wires to the backpressure transducer EGR solenoid come from the PCM and supply a pulse width modulated signal that will open and close the vacuum valve. The now pulse width modulated vacuum source is used to open the EGR valve. What is unique about this system is the monitoring of the quantity of exhaust flowing. Ford used a dedicated sensor. GM used the MAP signal. Chrysler uses the signal coming off of the O2Ss. The theory behind this is actually simple. If you add exhaust to the intake manifold, it will displace oxygen and the O2S signal will rise, looking like a rich condition. The opposite is also true; if you decrease the amount of exhaust in the intake, the cylinder will see more oxygen and the signal will drop or go lean. In this way the PCM looks at flow by knowing the amount of exhaust in the intake and looking for a corresponding change in the signal off of the O2S. Some vehicles will use the short-term fuel trim to indicate flow, which is also O2S driven. Don't forget, these systems require a functioning and accurate O2S. If, for some reason, the O2S monitor will not or cannot run, the EGR system should not run. If, as a technician, you are faced with an EGR monitor on a Chrysler that runs and an O2S that has not, you need to question the results. Something is preventing the O2S from running, and yet its signal, which may not be accurate, is being used to indicate EGR flow. There is an inconsistency that must be fixed. The O2S must run and not generate any DTCs, so that its signal is accurate enough to be used for other monitors. The order and sequence of monitors running is important and will be followed if the various components, especially the oxygen sensors, are functioning properly.

CONCLUSION

In this chapter, we took a detailed look at a frequent source of DTCs, the EGR system. We analyzed how bleeding in exhaust reduces the temperature of the burn and the level of NO_x produced within the cylinder. We examined different systems from both a control and measuring standpoint. We analyzed the use of a sensor to measure directly (DPFE) or a sensor to indirectly measure (MAP or O2S). We discussed the enabling criteria for EGR and identified some variables that might cause it to not run. In addition, we looked at the function of the primary sensor for EGR flow and noted its importance in various systems.

REVIEW QUESTIONS

1. Technician A states that the EGR helps to keep the combustion temperatures below 2500 degrees F. Technician B states that the EGR system will lower the development of NO_x if it is functioning correctly. Who is correct?

    a. Technician A only

    b. Technician B only

    c. Both Technician A and B

    d. Neither Technician A nor B

2. The use of a DPFE allows

    a. control of the flow of exhaust

    b. the measurement of the amount of exhaust flow

    c. an increase in the level of NO_x

    d. the monitor to run without following any enabling criteria

3. A GM EGR is opened for a 50% PWM. The MAP shows no change in vacuum. Technician A states that this indicates the EGR opened the required amount. Technician B states that the amount of flow is too much. Who is correct?

    a. Technician A only

    b. Technician B only

    c. Both Technician A and B

    d. Neither Technician A nor B

4. As the EGR flow is increased

    a. NO_x is increased

    b. cylinder temperature is increased

    c. cylinder temperature is decreased

    d. engine vacuum increases

5. A Ford DPFE is being discussed. Technician A states that the signal from the DPFE does not change as EGR flow is increased. Technician B states that the signal from the DPFE will rise as EGR flow occurs. Who is correct?

   a. Technician A only

   b. Technician B only

   c. Both Technician A and B

   d. Neither Technician A nor B

6. As EGR PWM is increased, the signal from an O2S should

   a. go lean

   b. go rich

   c. stay the same

   d. cause the short-term fuel trim to decrease

7. As EGR PWM is increased, the signal from the MAP should show

   a. increase in manifold vacuum

   b. pulse width signal also increase

   c. decrease in manifold vacuum

   d. a rich signal

8. The signal from an O2S is bad enough to set a DTC. Technician A states that this might shut down the EGR system and its monitor. Technician B states that fuel trim will override the O2S signal. Who is correct?

   a. Technician A only

   b. Technician B only

   c. Both Technician A and B

   d. Neither Technician A nor B

9. A P0401 is set. This indicates

   a. too much EGR flow

   b. a faulty EGR sensor

   c. a leak in the system

   d. too little EGR flow

10. Technician A states that the backpressure transducer measures exhaust pressure and determines the amount of EGR the engine might require. Technician B states that the enabling criteria for the EGR will require that conditions of load and speed be met to run. Who is correct?

    a. Technician A only

    b. Technician B only

    c. Both Technician A and B

    d. Neither Technician A nor B

AIR MONITOR, THERMOSTAT, AND PCV SYSTEMS

OBJECTIVES

At the conclusion of this chapter you should be able to: ■ Recognize the necessity of the AIR system on some vehicles ■ Understand the basic chemistry of adding additional O_2 to the exhaust stream ■ Follow the enabling criteria of an AIR-equipped vehicle ■ Follow a wiring diagram for an AIR-equipped vehicle ■ Diagnose an AIR system using a wiring diagram ■ Explain the function of the thermostat and understand the importance of engine temperature ■ Define a P0125 and a P0128 ■ Explain the function of the PCV system ■ Diagnose the PCV system based on DTCs

INTRODUCTION

In this chapter we look at some additional monitors. They are not found on all vehicles but will function like any other monitor. The first one is the AIR (air injection reaction) monitor. AIR as it is abbreviated is not a common system on vehicles of today. However, you may see it especially on some early OBD II vehicles. It is strictly an emission system responsible for adding additional air to the exhaust stream to improve catalytic converter operation. The second monitor we will discuss is probably the simplest one, the thermostat. This monitor showed up on vehicles in the early 2000s. We will also look at the PCV monitor, which is found on vehicles after 2002. It is important to note that there is a significant difference in these three monitors. The AIR system will have a specific monitor listed on the readiness page of your scanner and, like EVAP, EGR, CAT, etc., will have a specific set of enabling criteria. The thermostat and PCV will not show up on the readiness screen of your scanner. They will be monitored through the use of other monitors. The thermostat monitor will typically be a part of the comprehensive component monitor, and the PCV will usually be a part of the fuel monitor.

AIR REJECTION REACTION (AIR)

Let's first look at the AIR system and why it is needed. Then we will examine the monitor. When catalytic converters were first installed on vehicles, it was found that some engines under some conditions had insufficient air available for the CAT to function. Remember the basic function of the CAT is to convert HC (hydrocarbons) and CO (carbon monoxide) into CO_2 (carbon dioxide) and H_2O (water). This is done by the addition of oxygen or air (Figure 14–1).

Most OBD II systems use or get the air needed while the vehicle is running lean. Remember that the O2S will normally run the system first lean and then rich. This back and forth or oscillating of the air/fuel ratio will usually create sufficient oxygen for the CAT to function. However, in some vehicles, especially when cold, there is insufficient oxygen to allow complete conversion. On these vehicles air will be added through the use of an AIR system (Figure 14–2).

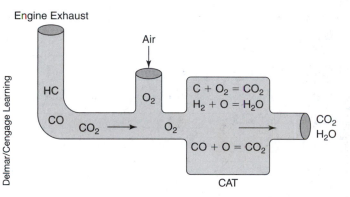

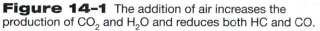

Figure 14-1 The addition of air increases the production of CO_2 and H_2O and reduces both HC and CO.

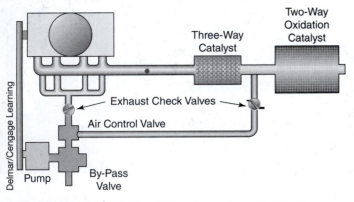

Figure 14-2 The AIR system injects O_2 into the exhaust system.

In this example an engine belt spins the pump, which pumps air into either the exhaust manifold or, in some cases, the CAT in between the three-way catalyst and the two-way catalyst. Many of the OBD II AIR systems will use an electrical pump rather than one that is belt driven. The system is controlled by the PCM so that it is functional only when needed. The vast majority of the OBD II vehicles that use AIR only use it under cold-engine conditions.

Notice that from the notes section (Figure 14-3) of the enabling criteria, the monitor begins to run after the system goes into closed loop. With heated O2Ss, closed loop is achieved within a few seconds to a minute. The engine idles and then drives to around 45 mph and the monitor runs. It is that simple. There are limited components that must be functional. The pump in Figure 14-4 will be either be spun eclectically or, on rare occasions, belt driven. Most systems now use an electric motor to spin the pump as it allows for more complete control by the PCM. Air is then pumped through some check valves (Figure 14-5).

The purpose of the check valves is to prevent high-pressure hot exhaust from reaching the pump, where it could be damaged. From here the air is pumped into the exhaust manifold through a series of tubes (Figure 14-6).

OBD II Drive Cycles	Oldsmobile
	Aurora

All Monitors	**1997**
Notes	1. If the Passive test, when equipped (Steps 1-3), fails to update the monitor status, the Active test, if required, will run in Step 4. The Active test begins after the engine goes into Closed Loop.
Conditions	1. Passive: ECT 39–221 degrees F., IAT 39–158 degrees F.; battery>11.7 volts. 2. Active: ECT 176–221 degrees F., IAT>50 degrees F.; battery>11.7 volts; engine in Closed Loop.
Step 1	Perform the I/M System Check. Failure to do so may result in difficulty in updating the status to YES.
Step 2	Turn OFF all accessories.
Step 3	Start the engine and idle for 2 minutes.
Step 4	Accelerate at part throttle to 45 mph and maintain speed for 3 minutes or until the monitor status updates to ready. If the AIR monitor status is YES, go to Step 12. If not, go to Step 5.
Step 5	Using a scan tool, access the DTC information. If there are any failed DTCs, diagnose and repair the DTCs.
Step 6	Determine which DTCs are required run in order to complete the test.
Step 7	Using a scan tool, observe the Not Ran Since Code Cleared display. Determine which DTCs that are required to have run, and have not run, for a YES status to be displayed.
Step 8	Enter the DTC in the specific DTC menu of the scan tool.
Step 9	Be sure the Enabling Conditions are met. Repeat the procedure until the scan tool indicates the diagnostic test has run.
Step 10	Repeat Steps 8-9 for any DTCs that have not run.
Step 11	Check the I/M System Readiness display. If the AIR system status is YES, go to Step 12. If not, refer to service manual.
Step 12	Check for emission-related DTCs. If any are present, go back to Step 5. If no DTCs are present, AIR system status is OK.

Figure 14-3 The enabling criteria for an AIR monitor.

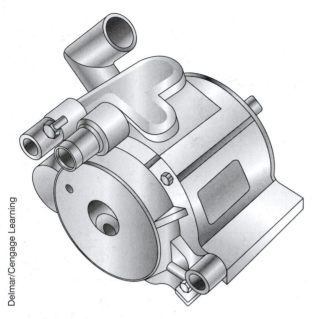

Delmar/Cengage Learning

Figure 14-4 The air pump in an AIR system may be electrically or belt driven.

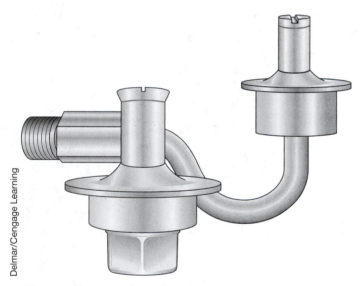

Delmar/Cengage Learning

Figure 14-5 Check valves prevent exhaust from flowing into the air pump.

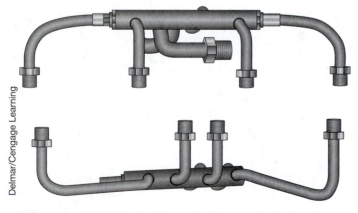

Delmar/Cengage Learning

Figure 14-6 Air tubes allow pressurized air to flow into the exhaust manifolds.

Let's look at the electrical diagram for an early GM OBD II AIR system (Figure 14–7). The system only has four components plus the usual power and ground circuits. Follow along as we detail how the system works. Power for the load side of the relay comes through the 50-amp AIR pump fuse, while power for the control side of the relay comes through the engine fuse, a 10-amp fuse in the underhood block. Both fuses are hot or powered in the run position of the ignition switch and feed both the relay and the secondary air injection solenoid. The negative side of the control circuit is opened and closed by the PCM. When the PCM recognizes the conditions that will require additional air, it will ground the relay coil, which will close the contacts. The closed contacts will bring power down to the AIR pump, which will begin to turn. Once pressure has been achieved, the PCM will energize the solenoid and air will flow through the system. Once air flows into the exhaust manifold, the O2S will see a lean condition and the PCM will "know" that the pump has functioned. As we previously stated, this is a very simple system capable of generating a few DTCs with easily achieved enabling criteria.

OTHER MONITORED COMPONENTS

If you look at a list of available monitors, you will see that beginning in 2000 manufacturers were required to add monitoring of the thermostat function and the positive crankcase ventilation (PCV). These two components are now monitored but not with their own separate monitor. The ability to generate DTCs is within the comprehensive component monitor. You will not see either of them listed on your scanner, but let's take a few minutes and discuss them.

The purpose of the thermostat monitor is to determine if the engine is generating sufficient heat, especially for catalytic function. While the vehicle is under a light load and moving faster than 15 mph, the PCM looks at the ECT and wants to see that the engine is warming up at a rate consistent with outside temperature (measured by the IAT). Typically the target temperature is a minimum of 20 degrees lower than thermostat opening temperature. With most thermostats opening at 195 degrees, the target temperature will be 175 degrees F (Figure 14–8).

Notice that the thermostat controls the operating temperature by shutting off the flow of coolant

Engine Controls Schematics (Secondary AIR Bypass Valve Solenoid, AIR Pump, IAC Valve and MAF Sensor)

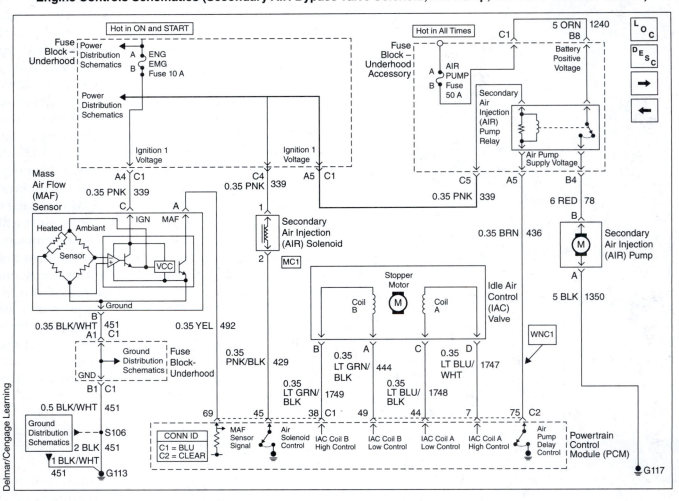

Figure 14-7 AIR system electrical wiring diagram.

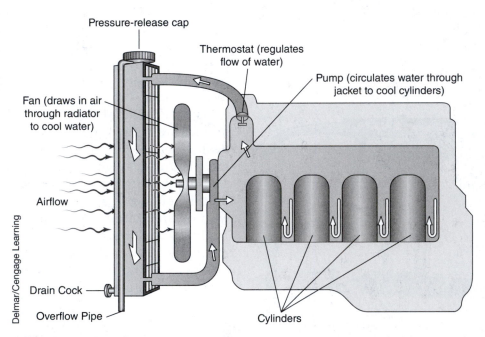

Figure 14-8 The thermostat controls coolant flow through the engine based on temperature.

to the radiator. Once the engine achieves minimum operating temperature, coolant is allowed to flow to the radiator, where it can be cooled by outside air. There are conditions, especially in colder climates, where at highway speeds the thermostat may not open at all. Under normal conditions and a cold start, the thermostat will close, trapping engine coolant. The engine heats up until the minimum operating temperature is achieved, and then the thermostat opens.

Early OBD II vehicles had a DTC (P0125) that was defined as "insufficient temperature for closed-loop test." On a vehicle with a thermostat monitor, an additional DTC is added (P0128). The definition of the P0128 is: "Coolant thermostat (coolant temperature below thermostat regulating temperature)." This DTC will turn on the MIL and save freeze frame just like other OBD II DTCs. This is an example of how the OBD II system has evolved since its inception in 1996.

POSITIVE CRANKCASE VENTILATION

The manufacturers also began monitoring the PCV system from model year 2002 on. Like the thermostat monitor, PCV does not have its own specific monitor and will not display readiness status. Figure 14–9 shows a simple diagram of a closed PCV system. "Closed" indicates that it is not open to the atmosphere.

The purpose of the system is to redirect any crankcase vapors into the intake system, where they will be burned with the normal mixture. It is an especially important system when an engine wears. Additional combustion gases leak past the piston rings where they can combine with the engine oil and cause additional wear. These gases contain HC and CO, which cannot be allowed to enter the atmosphere and add to the pollution problem.

The system is usually monitored by comparing the amount of air entering the engine by the mass air flow meter and doing the math required to figure out what the O2S signal should indicate at specific speed and load. If the numbers do not match, they indicate a leak, which may be within the PCV system. As with the thermostat, there are no specific enabling criteria and no specific monitor listed on your scanner. The PCV system is part of the fuel monitor and is capable of turning on the MIL just like another OBD II failure.

The system has become more complicated, as indicated by the 2010 Ford engine in Figure 14–10. The PCV valve is part of the oil seperator on the valve cover. Some PCV valves incorporate a small electrical heater that helps eliminate sludge buildup in low-temperature applications. Figure 14–11 shows a heated PCV on the left. Notice the electrical connection.

Positive Crankcase Ventilation System

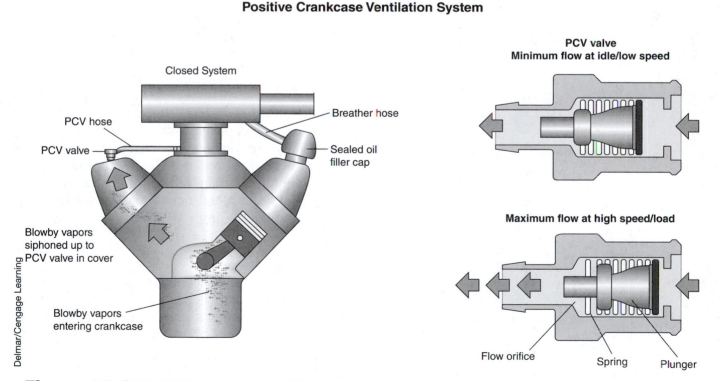

Figure 14-9 The PCV system helps to eliminate crankcase vapors.

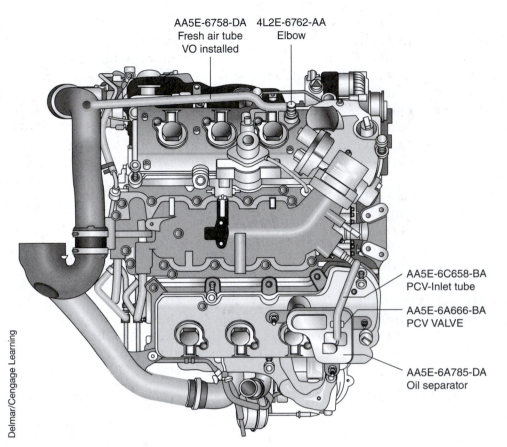

AA5E-6758-DA
Fresh air tube
VO installed

4L2E-6762-AA
Elbow

AA5E-6C658-BA
PCV-Inlet tube

AA5E-6A666-BA
PCV VALVE

AA5E-6A785-DA
Oil separator

Delmar/Cengage Learning

Figure 14-10 2010 Ford PCV system.

**Heater quarter-turn
PCV valve with heater**

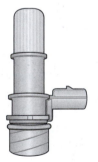

Assembled PCV system

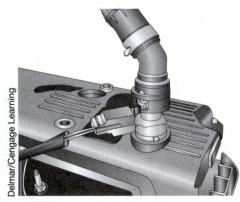

Delmar/Cengage Learning

Figure 14-11 PCV valves are
frequently installed in the valve cover.

On the right the PCV is installed in a valve cover assembly. The valve is designed to not require any service during the life of the vehicle.

WHAT DOES THE FUTURE HOLD?

OBD II is the system of today and the future. It should not come as a surprise that additional monitors will be added to the vehicle. Thermostat and PCV are good examples of the flexibility of the system. In 1996, when OBD II was introduced, the function of the PCV or the thermostat was not a concern; but by early 2000, monitoring of their function became standard. Additional systems and/or components will no doubt become part of the future OBD II.

CONCLUSION

In this chapter we looked at the AIR monitor and saw how it acts as an emissions device. We examined the chemistry of combustion, learned how the AIR system functions, and how it could be diagnosed using the enabling criteria and a wiring diagram. We looked at both the thermostat and PCV as they relate to OBD II and the setting of realted

DTCs. We also discussed recent changes in the monitoring system.

REVIEW QUESTIONS

1. Technician A states that the AIR system is needed on engines that have insufficient O2 for the CAT to be functional. Technician B states that the AIR system will decrease the amount of HC and CO coming out of the tailpipe. Who is correct?

    a. Technician A only

    b. Technician B only

    c. Both Technician A and B

    d. Neither Technician A nor B

2. When O_2 is added to the exhaust stream, tailpipe exhaust should contain mostly

    a. CO and CO_2

    b. HC and CO

    c. CO_2 and H_2O

    d. CO and H_2O

3. The enabling criteria for AIR is usually run

    a. just after the vehicle achieves closed loop

    b. when the vehicle cannot achieve closed loop

    c. when the vehicle is fully warmed up and at highway speeds

    d. at idle

4. According to the wiring diagram (Figure 14–7), power for the AIR pump comes through

    a. a relay

    b. the PCM

    c. the AIR solenoid

    d. none of the above

5. Technician A states that the AIR check valves are installed to protect the pump. Technician B states that the check valves prevent exhaust from flowing backwards into the system. Who is correct?

    a. Technician A only

    b. Technician B only

    c. Both Technician A and B

    d. Neither Technician A nor B

6. The thermostat monitor is capable of generating what DTC?

    a. P0300

    b. P0225

    c. P0128

    d. P1125

7. Technician A states that the thermostat monitor has its own PID on a scanner. Technician B states that the thermostat monitor is part of the comprehensive component monitor. Who is correct?

    a. Technician A only

    b. Technician B only

    c. Both Technician A and B

    d. Neither Technician A nor B

8. The purpose of the PCV system is to

    a. prevent excessive CO from being produced

    b. keep crankcase vapors from reaching the atmosphere

    c. allow the fuel monitor to run

    d. allow the CAT to function effectively

9. Two technicians are discussing the PCV system on a vehicle. Technician A states that the system is designed to re-burn crankcase vapors. Technician B states that the PCV system has its own set of enabling criteria that are sometimes difficult to run. Who is correct?

    a. Technician A only

    b. Technician B only

    c. Both Technician A and B

    d. Neither Technician A nor B

10. The PCV system may be monitored as part of which monitor?

    a. CAT

    b. Misfire

    c. EGR

    d. Fuel

USING OBD II MODES 5 AND 6

OBJECTIVES

At the conclusion of this chapter you should be able to: ■ Use a scanner and access operational modes ■ Match communications protocol to manufacturer's specifications ■ Access mode 5 and 6 specifications ■ Understand and be able to define TID and CID ■ Use a scanner and access data for specific TIDs and CIDs ■ Define substitute values ■ Use mode 5 to look at B1S1 and B2S1 values to determine O2S function ■ Determine if substitute values are present in mode 5 and/or 6 ■ Calculate actual sensor values off of mode 6 data

INTRODUCTION

As the use of OBD II as an emission test has increased, so has the incidence of reject status due to insufficient monitors run. Most State emission tests use the same standard for reject: two monitors allowed to be in a not-run state for vehicles produced from 1996–2000 and one monitor allowed for vehicles after 2001. With as many as seven monitors available, the pressure is on for the technician to set the monitors. Additionally, technicians have recognized that allowing a vehicle to leave the shop after a repair with the monitor in a not-run state is not a good idea. After the consumer has driven the vehicle for a period of time and a monitor runs and finds a problem, the check engine light comes back on and the customer begins questioning the repair shop. This is not a good situation to be in, but it can be avoided by using the enabling criteria and setting the monitors prior to giving the vehicle back to the owner. However, there are situations where a monitor must be run twice to be complete but will only run once. It usually takes the monitors running twice to determine a problem and set a DTC. If the monitor will not run to completion, there likely will not be any DTCs set to steer the technician to the problem. This is where mode 5 and/or mode 6 may be useful.

When you connect a scanner to an OBD II DLC, you will have the opportunity to determine what mode you wish to view. Figure 15–1 shows the choices. We spend the majority of our time looking at the first three modes because they contain the most

OBD II Operational Modes

- 1 = Datastream
- 2 = Freeze frame
- 3 = DTCs
- 4 = Clear DTCs and monitor status
- **5 = O2S data**
- **6 = Non-continuous monitor data**
- 7 = Continuous monitor data
- 8 = EVAP
- 9 = Vehicle information

Delmar/Cengage Learning

Figure 15–1 The nine modes that a scanner displays.

information. You can also see that when we clear codes we are using mode 4. Modes 5, 6, and 7, however, refer to data that is related to the monitors and their status. These three modes were originally designed for the manufacturer to check operation of various systems as the vehicle left the assembly plant. It was only after a few years that technicians found the monitor modes and the useful information they contained. Mode 7, which is continuous monitor data, is not very useful. The continuous monitors run so easily that looking at mode 7 data is rarely done.

Before we get into the use of the modes, there are a few ground rules we need to address. The first is the

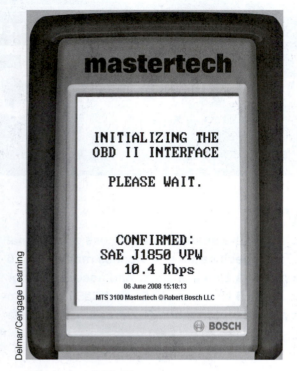

Figure 15-2 The scanner will identify the communications protocol being used by the vehicle.

simple fact that *not all scanners display mode 5 or 6.* Some inexpensive scanners will only show readiness, scan data, DTC's, and allow for the clearing of any DTC's. You will not see many of these used in repair shops because of their limited capability. There are also some manufacturer's scanners that will not display generic data, and since modes 5 and 6 are found on the generic side, they might not be available on a manufacturer's specific scanner. The last thing to remember is that all scanners need to be updated frequently. Using a scanner with outdated software can give you wrong information, which is actually worse than no information. Make sure the software is not out of date by more than a few years.

The second point that needs to be emphasized is that *frequently modes 5 and 6 are not easy to navigate through.* Remember that the original use for these modes was at the manufacturing level, where engineers had data that they needed. It was never designed for a technician's use. There will be times that you will look at the data and wonder what it means and why it is not in easily understood terminology. To be successful with these modes you will have to have the manufacturer's data specific to the vehicle you are working on and be capable of interpreting it. This brings up an interesting point. When the manufacturer transmits the data, it will follow a communications protocol. Your scanner will show you that protocol. Figure 15–2 shows a confirmed communication protocol of SAE J1850. Once we know this, we can look up data to make sure it has the same protocol. Figure 15–3 shows a listing of mode 6 data for use with SAE J1850 protocol. This is the only data list that can be used.

Don't worry about what the data is representing at this point. The important line is the title where J1850 is listed. The data definitions list *must* match

Mode $06 data definitions for GM vehicles using J1850/Class2 diagnostic data link
Some items have footnotes, defined on the last pages.

Test ID (Hex)	Comp ID (Hex)	Test Limit Type	Description (see footnotes on last page)	Decimal Range	Hex Range
			**Secondary Air Injection Reaction System Monitor**		
03	01	0-High	AIR bank 1 test	0 -to- 65535 counts	0000 - FFFF
03	02	0-High	AIR bank 2 test	0 -to- 65535 counts	0000 - FFFF
03	03	0-High	AIR on pressure error test bank 1 Out of range high	−32.768 -to- +32.767 kPa	0000 - FFFF
03	03	1-Low	AIR on pressure error test bank 1 Out of range low	−32.768 -to- +32.767 kPa	0000 - FFFF
03	04	1-Low	AIR valve shut pressure error test bank 1	−32.768 -to- +32.767 kPa	0000 - FFFF
03	05	0-High	AIR pump off pressure error test bank 1	−32.768 -to- +32.767 kPa	0000 - FFFF
03	13	0-High	AIR on pressure error test bank 2 Out of range high	−32.768 -to- +32.767 kPa	0000 - FFFF
03	13	1-Low	AIR on pressure error test bank 2 Out of range low	−32.768 -to- +32.767 kPa	0000 - FFFF
03	14	1-Low	AIR valve shut pressure error test bank 2	−32.768 -to- +32.767 kPa	0000 - FFFF
03	15	0-High	AIR pump off pressure error test bank 2	−32.768 -to- +32.767 kPa	0000 - FFFF
03	16	0-High	AIR on pressure differential high between bank 1 and bank 2 Out of range high	−32.768 -to- +32.767 kPa	0000 - FFFF
03	16	1-Low	AIR on pressure differential high between bank 1 and bank 2 Out of range low	−32.768 -to- +32.767 kPa	0000 - FFFF

Figure 15-3 Data definitions must match the protocol in use.

Figure 15-4 An ISO 14230-4 protocol confirmed.

the communications protocol shown on the scanner, or the data is worthless. Figure 15–4 shows a different vehicle with a different protocol.

Trying to use the J1850 definition list with this ISO 14230-4 protocol will result in errors that may take the technician off into some areas that look bad when they really are not. Always match protocol!

The last part of the ground rules is that *there may be data on the scanner for unused section of the monitor.* For example, a manufacturer uses an EGR solenoid on some of its vehicles but not on all. It is likely that a vehicle without the solenoid will show data and probably a fail for the EGR solenoid. This makes sense if you think about it. If the test (a part of the monitor) is looking for the function of the EGR solenoid and does not

```
MTS 3100 Mastertech

TID$02 CID$04 ···· PASS
TID$02 CID$26 ···· PASS
TID$02 CID$36 ···· PASS
TID$02 CID$46 ···· FAIL
TID$02 CID$10 ···· PASS
TID$02 CID$20 ···· PASS
TID$02 CID$30 ···· PASS
TID$02 CID$40 ···· PASS
TID$02 CID$12 ···· PASS
TID$02 CID$11 ···· PASS
TID$02 CID$21 ···· PASS
TID$02 CID$31 ···· FAIL

25 June 2005 13:32:32
MTS 3100 Mastertech © Vetronix Corporation
```

Figure 15-5 The scanner is displacing each CID's status for a single monitor.

see it, it fails. You have to be able to recognize that this fail indicates a test for a component that is not on the vehicle. It is the manufacturer's data definition list that will show this. Figure 15–5 shows an example.

Notice that there are two fails listed. Let's look at the top one, which is identified as TID$02 CID$46. Don't worry about TID (test identification) and CID (component identification) at this point. If we take the failed test and try to find out what is being tested by looking at the manufacturer's specification, we can see the data definitions as in Figure 15–6. These definitions can be useful in directing a technician to the mode 5 or 6 data that is failing.

Enhanced Evaporative Emission System Monitor #1 (.040" Leak)		
02	12	EVAP small leak test
02	20	EVAP weak vacuum fail test 1
02	21	EVAP purge leak vapor fail test
02	21	EVAP purge leak vacuum fail test
02	26	EVAP excess vacuum test 1
02	36	EVAP excess vacuum fail test 2
02	52	EVAP small leak test
02	60	EVAP weak vacuum fail test 1
02	62	EVAP NV .020" Error Test
02	66	EVAP excess vacuum test 1
02	71	EVAP purge leak vacuum fail test
02	72	EVAP NV .040" Error Test
02	84	EVAP cannister loading test
02	86	EVAP excess vacuum pass test 2
02	90	EVAP weak vacuum pass test 1
02	91	EVAP purge leak pass test
02	B0	EVAP weak vacuum test 2 vacuum
02	C0	EVAP weak vacuum test 2 vapor
02	C6	EVAP excess vacuum pass test 2
02	D0	EVAP weak vacuum pass test 1

Figure 15-6 The data definitions for a single monitor.

The TIDs are listed in the first column, and the CIDs are listed in the second. Our fail was for TID$02 and CID$46, which are not on the list. This test failed because the component being tested is not on the vehicle. This is another reason why the communications protocol shown on the scanner must match that shown on the data definition list. If, in our example, you were looking on the wrong protocol, there might be a $46 listed. If you have a fail for a listed mode 5 or 6 test, you may waste time searching for a problem that is not on the vehicle.

The last point to make regarding ground rules is more of a suggestion than a rule. You will have to *practice using your scanner in modes 5 and 6* to get good and reasonably fast. No one will pick up a scanner and begin making money by diagnosing problems using mode 5 and 6 immediately. It is impossible to follow through these modes and have it all make sense. Practice, practice, practice: You will only get good through time on task. Use these modes on vehicles that do not have anything wrong with them so you see how your scanner will function. Then, when a vehicle arrives that will not run the monitors, your practice will pay off.

Let's practice on a vehicle without problems and walk through the use of mode 6. First some definitions; a TID is a test identification. It is another way of saying a monitor. Generally, a single monitor will have one TID. For instance, our prior example had a TID$02. All of the 02s on this vehicle are for EVAP testing. How about the $? This is an indication that the number system is hexadecimal rather than base 10 like we are used to. It is frequently an engineering system, which will use a number system that has a base of 16 rather than the ten we normally use. Additionally, it will use numbers 0–9 and letters A through F. This will allow the scanner to display numbers from 1 to 256, which are computer numbers. Remember when you see a number that is a computer number—for example, 32, 64, 128, 256, 512, 1024, etc.—that you are probably looking at a substitute value. Substitute values are put into the system by the PCM prior to running the monitors or sometimes when the monitor will not run to completion. We will look at substitute values later in this chapter.

The next definition is for CID or component identification. This is a bit misleading since most of us are used to a component being something we can hold in our hand, like a coolant temperature sensor or a mass airflow. A CID is a piece or component of a monitor. Look at Figure 15–7. It shows all of the pieces or components of the oxygen monitor.

Test ID (Hex)	Comp ID (Hex)	Test Limit Type	Description (see footnotes on last page)	Decimal Range	Hex Range
	**Oxygen Sensor Monitors and Constants**				
05	01	0-High	Rich to lean sensor threshold voltage - B1S1	0 -to- 2048 mV	0000 - FFFF
05	02	0-High	Lean to rich sensor threshold voltage - B1S1	0 -to- 2048 mV	0000 - FFFF
05	03	0-High	Low sensor voltage for switch time calculation - B1S1	0 -to- 2048 mV	0000 - FFFF
05	04	0-High	High sensor voltage for switch time calculation - B1S1	0 -to- 2048 mV	0000 - FFFF
05	05	0-High	Rich to lean sensor switch time - B1S1 Out of range high	0 -to- 1024 ms	0000 - FFFF
05	05	1-Low	Rich to lean sensor switch time - B1S1 Out of range low	0 -to- 1024 ms	0000 - FFFF
05	06	0-High	Lean to rich sensor switch time - B1S1 Out of range high	0 -to- 1024 ms	0000 - FFFF
05	06	1-Low	Lean to rich sensor switch time - B1S1 Out of range low	0 -to- 1024 ms	0000 - FFFF
05	07	1-Low	Rich to lean switches - B1S1	0 -to- 65535 switches	0000 - FFFF
05	08	1-Low	Lean to rich switches - B1S1	0 -to- 65535 switches	0000 - FFFF
05	09	0-High	R/L response to L/R response ratio - B1S1 Out of range high	0:1 -to- 8:1 ratio	0000 - FFFF
05	09	1-Low	R/L response to L/R response ratio - B1S1 Out of range low	0:1 -to- 8:1 ratio	0000 - FFFF
05	0A	0-High	Post catalyst sensor open test - B1S2 [4]	0 -to- 65535 samples	0000 - FFFF
05	0B	1-Low	Post catalyst sensor rich tests - B1S2	0 -to- 2048 mV	0000 - FFFF
05	0C	0-High	Post catalyst sensor lean tests - B1S2	0 -to- 2048 mV	0000 - FFFF
05	0D	0-High	Difference between R/L response and L/R response - B1S1 Out of range high [2]	−32768 -to- +32767 ms	0000 - FFFF
05	0D	1-Low	Difference between R/L response and L/R response - B1S1 Out of range low [2]	−32768 -to- +32767 ms	0000 - FFFF
05	13	0-High	Low sensor voltage for half period time calculation - B1S1	0 -to- 2048 mV	0000 - FFFF
05	14	0-High	High sensor voltage for half period time calculation - B1S1	0 -to- 2048 mV	0000 - FFFF

Delmar/Cengage Learning

Figure 15-7 The O2S monitor ($05) for this vehicle contains 19 pieces or components.

Notice that all components of the oxygen sensor monitoring carry the TID of $05, but there are 19 CIDs associated with the monitor. It is hard to tell from the list if any of the CIDs are more important than others, but a priority list exists within the PCM. It is this priority list that can sometimes stop monitor operations after one running. The key to understanding the importance of modes 5 and 6 is knowing how the system functions. The monitor will probably need to run twice to set DTCs, and yet it might shut down the second running of the monitor if it sees something important (that is not correct) during the first running of the monitor. If you are only looking at the monitor status screen, as Figure 15–8 shows, you cannot tell if the monitor even ran once.

Basically, the monitor status screen is only telling you half of the necessary information. It is saying that the monitors have not run to completion. To run to completion means to run twice, not just once. It is at this point that you would switch over to mode 5 or 6 for more information.

Let's go through some mode 5 and 6 examples so you can see how it might be useful. To be successful you have to do the same thing with your scanner. Practice, until it is second nature. Don't forget that modes 5 and 6 are on the generic side of the PCM. After connecting your scanner, turn the ignition key on and turn the scanner on. Your screen should appear similar to Figure 15–9.

Global is another way of saying generic. By choosing global you are accessing data on the generic side of the PCM. Once we select global the scanner offers additional choices (Figure 15–10).

By selecting F1 we are choosing OBD II, and a functions list should appear on the screen, as in Figure 15–11.

This functions list is similar to the mode list that we looked at in the beginning of this chapter. F5 will take us into system testing, which is where we want to go. After selecting F5, the screen should allow us into either mode 5 (O2S testing) or mode 6 (other), as shown in Figure 15–12.

Figure 15-9 Mode 5 and/or 6 is accessed from the generic or global side of the PCM.

Figure 15-8 A list of those monitors supplied and whether they have run to completion or not.

Figure 15-10 Available data on the generic side of the PCM.

Figure 15-11 System tests will contain most PCM tests.

Figure 15-12 Mode 5 will be displayed in O2S test results. Mode 6 will be in other results.

F2 should get us directly into mode 6 (Figure 15–13). By selecting it, the screen shows a list of various CIDs that are all dealing with the same series of tests. We know this because the TIDs for all lines are all #02s.

Figure 15-13 A list that shows two failed CID tests.

A check of the manufacturer's information reveals that on this vehicle all $02s are EVAP oriented (Figure 15–14).

All $02s are listed for leak detection of a large leak (.040" or greater). This information was found on the manufacture's website. Without it, we might not have access to the very important definitions for the TIDs.

If we go back to the master list and choose one of the TIDs that was a pass (CID $20), we see Figure 15–15.

Our definition lists CID $02 and TID $20 as "EVAP weak vacuum check." If, during the monitor running, this shows as a fail, it would indicate that the vacuum level is lower than specification. Follow along as we define the rest of the information on the screen. Supporting ECU is $10. This is the engine PCM. Minimum value shows as N/A, meaning that no real value has been stated. This is frequently the case if the monitor has only run once. Because no minimum value is specified, it also means that the system will show increasing numbers as the vacuum level drops. Once the vacuum level has dropped too far to be functional, the Cur Val will exceed 500. Cur Val stands for current value, and our vehicle is running a 5. With a Max Val of 500 and no value for min, you could think of this as a range between 0 and 500, with lower numbers being better. A 5 is just about perfect. This is why TID $02 and TID $20 are a pass. At this point, some scanners can give you additional data regarding what definitions the CID and TID are.

Let's follow the same logic with a fail. The scanner shows a fail for TID $20 CID $46 on the master list, and

Enhanced Evaporative Emission System Monitor #1 (.040" Leak)		
02	12	EVAP small leak test
02	20	EVAP weak vacuum fail test 1
02	21	EVAP purge leak vapor fail test
02	21	EVAP purge leak vacuum fail test
02	26	EVAP excess vacuum test 1
02	36	EVAP excess vacuum fail test 2
02	52	EVAP small leak test
02	60	EVAP weak vacuum fail test 1
02	62	EVAP NV .020" Error Test
02	66	EVAP excess vacuum test 1
02	71	EVAP purge leak vacuum fail test
02	72	EVAP NV .040" Error Test
02	84	EVAP cannister loading test
02	86	EVAP excess vacuum pass test 2
02	90	EVAP weak vacuum pass test 1
02	91	EVAP purge leak pass test
02	B0	EVAP weak vacuum test 2 vacuum
02	C0	EVAP weak vacuum test 2 vapor
02	C6	EVAP excess vacuum pass test 2
02	D0	EVAP weak vacuum pass test 1

Delmar/Cengage Learning

Figure 15-14 The data definition list for 0.40" leak testing of an EVAP system.

Delmar/Cengage Learning

Figure 15-15 The current value of 5 with a maximum value of 500 indicates a pass.

Figure 15-16 A current value of 0 with a minimum value of 100 indicates a fail.

obviously a fail is not good. By looking at the parameters of the fail, we should be able to determine why it is listed as a fail. Figure 15–16 shows the scanner data.

The data looks the same as the pass, except notice the Min Val is 100 and there is N/A listed for Max Val. This means that the values will move down until they are lower than 100. At this point a fail would be listed. With a current value of 0, which is lower than 100, it is obvious why it is listed as a fail.

Now let's go back to the definition list to find the $46 (Figure 15–17).

Notice that the numbers run from 36 and jump to 52. There is no 46. This component ($46) of the EVAP ($02) test apparently does not exist on this vehicle. The importance of this cannot be overstated: You must have manufacturer information and use it to direct you through the process of using mode 6.

Let's jump to the data available within mode 5. Remember mode 5 is for the oxygen sensors. Sometimes mode 5 data will be a bit easier to follow because it is frequently in a format that we automatically recognize. You have seen that mode 6

Enhanced Evaporative Emission System Monitor #1 (.040" Leak)			
02	12		EVAP small leak test
02	20		EVAP weak vacuum fail test 1
02	21		EVAP purge leak vapor fail test
02	21		EVAP purge leak vacuum fail test
02	26		EVAP excess vacuum test 1
02	36		EVAP excess vacuum fail test 2
02	52		EVAP small leak test
02	60		EVAP weak vacuum fail test 1
02	62		EVAP NV .020" Error Test
02	66		EVAP excess vacuum test 1
02	71		EVAP purge leak vacuum fail test
02	72		EVAP NV .040" Error Test
02	84		EVAP cannister loading test
02	86		EVAP excess vacuum pass test 2
02	90		EVAP weak vacuum pass test 1
02	91		EVAP purge leak pass test
02	B0		EVAP weak vacuum test 2 vacuum
02	C0		EVAP weak vacuum test 2 vapor
02	C6		EVAP excess vacuum pass test 2
02	D0		EVAP weak vacuum pass test 1

Delmar/Cengage Learning

Figure 15-17 The data definition list indicates which tests are run on specific vehicles.

Delmar/Cengage Learning

Figure 15-18 Oxygen sensor testing usually has a separate results group.

requires a manufacturer's list to make sense of it. If we go back to the two choices within OBD II generic, we see Figure 15–18.

F1: O2S TEST RESULTS looks like the place to be. When it is selected, we see two screens of data for the two front oxygen sensors (Figure 15–19).

B1S1 is on the left, and B2S1 is on the right. Notice that there are actual voltages that we are familiar with and times for rich to lean and lean to rich transitions. The first four lines of data are actually the specifications for the tests. The bottom of each screen contains three TIDs, but the critical information is contained within the rich to lean line (R>L SW TIM). This is the time in seconds that the

O2S takes to switch from full rich to full lean. The next line (L>R SW TIM) is the time in seconds from full lean to full rich. Look at both of these sensors. Which one is the slowest? B1S1 is at 0.028sec and 0.020sec, while B2S1 is at 0.016 and 0.012. B1S1 is substantially slower, and we have seen how O2S speeds can prevent monitors from running. These scans are from a vehicle that will not run the monitors to completion. Modes 5 and 6 revealed that the monitor had run only once and not the second time required to run to completion. After many hours of diagnostics, the technician noticed that B1S1 was slower that B2S1. After replacing the B1S1 O2S, all monitors ran within 20 minutes of driving the enabling criteria.

Why wasn't the check engine light on and a DTC stored for this bad O2S? Hopefully, by now, you realize that the monitors generally have to run twice to set DTCs, and this vehicle would only run once. The lack of speed prevented the O2S monitor from running the second time and setting the DTC. The signals from functioning O2Ss would have been used for other monitors, but because the O2S would not run to completion, the rest of the OBD II system was effectively shut down. This vehicle was rejected five times at an emissions testing facility, and had been to three shops. The consumer had already spent over $1,000.00 on the vehicle trying to get the monitors to run. It took a talented, up-to-date technician familiar with modes 5 and 6 to figure this one out.

```
MTS 3100 Mastertech

R»L O2S V·········0.495V
L»R O2S V·········0.495V
LOW SW V··········0.295V
HIGH SW V·········0.595V
R»L SW TIM·0.028sec
L»R SW TIM·0.020sec
TID $70···········236CNT
TID $71···········238CNT
TID $81················39


25 June 2005 13:40:09
MTS 3100 Mastertech © Vetronix Corporation
```

```
MTS 3100 Mastertech

R»L O2S V·········0.495V
L»R O2S V·········0.495V
LOW SW V··········0.295V
HIGH SW V·········0.595V
R»L SW TIM·0.016sec
L»R SW TIM·0.012sec
TID $70···········255CNT
TID $71···········255CNT
TID $81················45


25 June 2005 13:41:55
MTS 3100 Mastertech © Vetronix Corporation
```

Figure 15-19 B1S1 (on left) O2S was substantially slower and resulted in shutting down the monitor.

SUBSTITUTE VALUES

This is a good time to discuss substitute values. Manufacturers program PCMs with the ability to test individual components, through the process of running monitors. However, prior to the actual running of the monitor there will be values displayed in modes 5 and or 6 designed to allow the vehicle to run from point A to point B. Remember that it takes the monitor running to completion to set DTCs and turn on the MIL. This means that if the monitor will not run to completion, substitute

values will not turn on the MIL or set any DTC. As a technician you will be able to recognize a substitute value either by tracking the value after the PCM is cleared or by realizing that the value is a computer value.

Let's look at the tracking of values first. If a vehicle arrives at the repair facility with monitors not run, a check of values within modes 5 or 6 should be done and the values written down. The PCM is then reset or cleared just as we do when clearing DTCs. A key on–engine off recheck of the values will now reveal the substitute values, and they should also be written down. Any change in the values from this point on will take place once the monitor has run. The change from a substitute value to a new value would indicate that the monitor has run at least once.

Frequently, substitute values will be computer numbers. A computer number is one that started with 4 and doubled: 4, 8, 16, 32, 64, 128, 256, 512, 1024, etc. If you see values that are computer numbers, you are likely looking at substitute values (Figure 15-20).

Notice that the Min Val is 512 and the Cur Val is 1024. Both of these numbers indicate that the computer has put substitute values in place until the monitor runs. Once the monitor runs, the values will no longer be substitute numbers and will change. The vehicle in Figure 15-20 came into the shop after numerous rejects. The customer had been repairing his own vehicle without the correct tools. After realizing that there was a problem, the technician repaired the faulty circuit and drove the enabling criteria, resulting in the scan shown in Figure 15-21.

Notice that both values have now changed. The previous 512 has changed to a 461, and the 1024 has changed to a 768. Ask yourself, what has occurred here? Hopefully the answer is the monitor ran and

```
MTS 3100 Mastertech

OBD II PARAMETER HELP

PARAMETER NAME:
  On-Board Monitoring for
  Non Continuous tests

ID.:   1, $(01) CID $11
SUPPORTING ECU'S:
  $10

REPORTING ECU:
  $10 (Engine)
Min Val: 512
Max Val: N/A
Cur Val: 1024
```

Figure 15-20 Substitute values that are computer values indicate that the monitor has not run.

Delmar/Cengage Learning

Figure 15-21 Real values appear once the monitor runs.

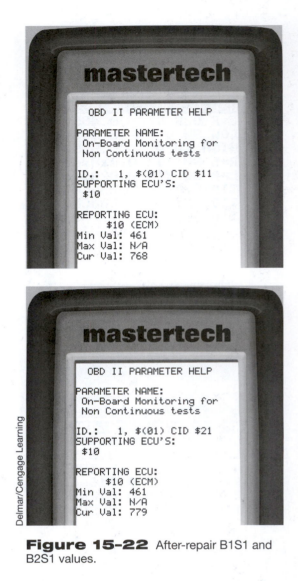

Figure 15-22 After-repair B1S1 and B2S1 values.

"real" values have replaced the substitute values. This circuit issue, once corrected, allowed the vehicle to pass the emission test, and yet during all of the rejects never turned on the MIL or set a DTC that could be used to guide the technician to the problem. A good technician took hours to track down this problem and used mode 6 as a road map.

The other very important point regarding this vehicle was why didn't the monitor run in the first place? It actually came down to the current value being so far apart on two O2Ss (B1S1 and B2S1) that the PCM shut down the monitor and prevented it from running. Once the circuit problem on the one O2S was corrected, the current values were now close enough that the monitor ran. Figure 15–22 shows a comparison of the two O2Ss after repair.

Notice that the current value for each is very close (768 and 779). Also notice that the Min Val is the same for both. This makes sense if you think about it. The Min Val is a value that the PCM is using as a benchmark, and B1 should match or be very close to B2.

You might be thinking it would be helpful if the engineering values in mode 5 or 6 could be converted to some value that makes sense. For the most part the values are just that—values. However a few manufacturers make available conversion formulas that may be useful. The two O2Ss in Figure 15–22 were from Ford products. If we look up mode 5 and 6 data, we find that Ford has conversion numbers shown in Figure 15–23.

Notice that $01 $11 and $01 $21 are both for voltage amplitude and voltage threshold but for different O2Ss. The bottom of the chart indicates that the voltage amplitude can be calculated by multiplying the Cur Val by 0.00098. B1S1 Cur Val is 768. $768 \times 0.00098 = .752$

or 752 mV. This appears to be the voltage threshold or maximum voltage. B2S1 Cur Val is 779. $779 \times 0.00098 = .763$ or 763 mV. In this case, the value is not as important as the difference between the voltages. 752 mV and 763 mV are close enough that the monitors ran. However, let's back up and compare the voltages when the monitor would not run. We started this section off with a B1S1 Cur Val of 1024, which we indicated was a substitute value. If we apply the same math to 1024 ($1024 \times 0.00098 = 1.00$), 1.00 V is the result, and this is a long way from 763 mV. It was the distance between the two values that prevented the monitor from running, not the fact that it was a substitute value.

Not all manufacturers give us conversion numbers, and realistically you will not need to do the math very frequently. The Cur Val number is what is important, and in the case of oxygen sensors it needs to be close or the monitor may not run.

J1979 Mode $06 Data			
Test ID	Comp ID	Description	Units
$01	$11	HO2S11 voltage amplitude and voltage threshold	volts
$01	$21	HO2S21 voltage amplitude and voltage threshold	volts
$03	$01	Upstream 02 sensor switch-point voltage	volts
Conversion for Test IDs $01 through $03: multiply by 0.00098 to get volts			

Figure 15-23 Sometimes mode 5/6 values can be converted to voltage.

CONCLUSION

Modes 5 and 6 are not just for any technician. You have to have a solid foundation in OBD II and understand the monitor process completely. These modes will not be the first place you go with monitors that will not run because using them is extremely time consuming. It is best to use a pencil and a large piece of paper as your diagnostic tools. Write down values when the vehicle arrives, clear the PCM, and begin running the enabling criteria until the values change. Stop the vehicle and write down the values again. A change indicates that the monitor has run once. This second set of values is where the road map to success might be.

REVIEW QUESTIONS

1. Technician A states that the PCM's operational modes will hold information regarding monitor status. Technician B states that operational modes are part of the generic side of OBD II. Who is correct?
   a. Technician A only
   b. Technician B only
   c. Both Technician A and B
   d. Neither Technician A nor B

2. The communications protocol is
   a. the method used by the manufacturer to generate DTCs
   b. the PCM's internal clock circuit
   c. the method of transferring data to a scanner
   d. an operational mode

3. Mode 5 or mode 6 has been accessed, and the main pass/fail screen is visible. Technician A states that the TIDs are the actual monitors in number format. Technician B states that the CIDs are the enabling criteria for the monitors. Who is correct?
   a. Technician A only
   b. Technician B only
   c. Both Technician A and B
   d. Neither Technician A nor B

4. Technician A states that a TID is a part of a monitor test. Technician B states that a CID is the entire monitor. Who is correct?
   a. Technician A only
   b. Technician B only
   c. Both Technician A and B
   d. Neither Technician A nor B

5. Data that indicates monitor run status is available
   a. only on the manufacturer's side of the PCM
   b. by looking at DTCs that have been retrieved
   c. by looking at the enabling criteria
   d. on the generic side as an operational mode

6. Technician A states that a substitute value is a value programmed into the PCM prior to the monitor running to completion. Technician B states that frequently substitute values will be computer numbers. Who is correct?
   a. Technician A only
   b. Technician B only
   c. Both Technician A and B
   d. Neither Technician A nor B

7. A scanner shows a CID as having a Min Val of 235, a Max Val of 2674, and a Cur Val of 559. This indicates that
   a. the "piece" of the monitor that ran passed
   b. the "piece" of the monitor that ran failed
   c. the monitor has not run
   d. a DTC is present in a different operational mode

8. A scanner shows a CID of no Min Val, a 1024 Max Val, and a Cur Val of 256. Technician A states that this indicates substitute values. Technician B states that this indicates the monitor has not run to completion. Who is correct?
   a. Technician A only
   b. Technician B only
   c. Both Technician A and B
   d. Neither Technician A nor B

9. A current value is displayed within mode 5 for an O2S. The value
   a. indicates that the monitor has run
   b. may be able to be converted to a real voltage
   c. will indicate whether the CID is listed as a pass or a fail
   d. all of the above

10. Two technicians are discussing modes 5 and 6. Technician A states that the values present will become substitute values after a DTC is cleared. Technician B states that the values present will change to real test values once a monitor has run. Who is correct?
    a. Technician A only
    b. Technician B only
    c. Both Technician A and B
    d. Neither Technician A nor B

REPAIR BASELINE TECHNIQUES

OBJECTIVES

At the conclusion of this chapter you should be able to: ■ Define and use various baseline techniques ■ Use the graph data function of a scanner as a baseline ■ Use the bi-directional data function to control an output device ■ Use bi-directional data to follow up on a repair involving a DTC prior to setting monitors ■ Follow a recognized procedure when diagnosing a vehicle ■ Explain short-term fuel trim and its relationship to long-term fuel trim ■ Explain long-term fuel trim and its relationship to short-term fuel trim ■ Use STFT and LTFT data PIDs to determine if a vehicle has been repaired ■ Use the monitor system to baseline a vehicle

INTRODUCTION

Before we get into various techniques, we need to have a working definition of a baseline technique. Simplified, it is how you will know that the vehicle has been repaired. OBD II is rather simple: run the monitors and have the MIL stay off. The combination of the monitors running and the MIL remaining off is the baseline. You will know the vehicle has been repaired once the MIL stays off. If it comes on as the monitors run, the vehicle is not repaired. OBD II allows the system to be the ultimate authority that the vehicle has been repaired. This is the function of the monitors. They will indicate that the vehicle is repaired, but only if setting all of them becomes a part of every job. The repair shop that allows the vehicle to leave without running the monitors is not doing the entire job. They are missing the baseline that will allow them to release the vehicle to the customer confident that it has been repaired. The unfortunate scenario repeated frequently is a repair facility fixes a vehicle, clears the DTC and the MIL, and allows the customer to take the vehicle. The shop is confident that they have fixed the vehicle but did not run the monitors to prove the job was complete. Three days later the monitors run and set another DTC or, worse yet, the same DTC that brought the vehicle into the shop in the first place. Consider what the customer is thinking at this point. The check engine light came back on after a short period

of time, so the vehicle has the same problem. He will either bring the vehicle back to the previous repair facility and argue that his vehicle was not repaired or take it somewhere else. Either way, the repair facility loses out.

The correct procedure starts with a scan of the monitor status when the vehicle is first brought into the repair facility. Let's use the Wisconsin EPA Vehicle Inspection Report or VIR as it is called (Figure 16–1).

Under the individual test results we can see that the CAT. EVAP, EGR, and O2S systems have not run. If this vehicle came in with a P0300 (random misfire), the shop needs to explain to the customer that they will fix the misfire, run the monitors, and repair any additional problems that the monitors reveal. Because the vehicle has not run most or all of the non-continuous monitors, there is no assurance that the misfire DTC is the only problem. There could be an EVAP leak, an EGR blockage, or a weak CAT that the customer is not aware of. The monitors were suspended when the misfire DTC set, so they would be incapable of detecting additional problems. If the shop just repairs the misfire DTC and clears it, the system will begin running the monitors and pick out the other problems. When the MIL comes back on, the customer has no idea that he now has an EVAP problem. The light was on last week and now it is back on, so in his mind, the original problem is back or was never

MVD-2470 - 2001

WISCONSIN VEHICLE INSPECTION PROGRAM

State of Wisconsin
Vehicle Inspection Report

9893269

Thank you for helping to clean up our air. Your vehicle's inspection was performed in accordance with applicable federal and state regulations. Your vehicle's inspection result is shown below in the block titled "Final Result."

Pass: If your vehicle's final result is PASS, please renew your registration as indicated on your renewal notice. Your inspection record is sent by computer to the Division of Motor Vehicles. Please do not mail this document with your renewal notice.

Fail: After repairs or adjustments are made, you are entitled to a reinspection (up to 2 reinspections are allowed). For the reinspection, drive your vehicle back to any Wisconsin emissions inspection station and present this report with the Repair Data section on the reverse side completed. Repairs for your vehicle may be covered by the manufacturer's emissions performance or defect warranty. Check your owner's manual or contact your dealer for details. **Retest requirements do not change your registration deadline.** DO NOT MAIL THIS DOCUMENT WITH YOUR RENEWAL NOTICE

IF YOU CHOOSE TO RENEW YOUR REGISTRATION IN PERSON, PLEASE HAVE THIS DOCUMENT WITH YOU.

INDIVIDUAL TEST SUMMARY

Emissions	Evaporative System	On-board Computer	Equipment	Recall	Final Result
					REJECT

VEHICLE INFORMATION

Plate	Plate Type	VIN	Vehicle Year	Make	Odometer in thousands
168PCW	AUT	2FAFP71W11X190675	2001	FORD	137

TEST INFORMATION

Date	Test Start Time	Test End Time	Station	Lane
17-APR-2009	16:42:03	16:44:58	05	3

INDIVIDUAL TEST RESULTS

	Reading	Units	Limit	Result
On-Board Computer: Unable to test the vehicle - following readiness monitors were not set: Catalyst Evaporative system Oxygen sensor Oxygen sensor heater EGR system				REJECT

Delmar/Cengage Learning

Figure 16-1 Vehicle inspection reports will usually show which monitors have not run to completion.

repaired. The correct procedure would have been to scan for monitor status when the vehicle first arrived at the repair facility. Then the technician could inform the customer that the correct procedure would be to fix the DTC, run the monitors, fix any additional DTC that set, and run the monitors again. The goal is to run the monitors to completion and not generate any additional DTCs. Then the vehicle is fixed.

USING THE GRAPH DATA FUNCTION

Most scanners in use in modern facilities have a variety of functions. One of these is the graph data function, which allows the technician to see a PID in a graph format and can be a wonderful baseline. Let's take an example of a vehicle with the MIL on and a CAT code (P0420). Prior to selling the customer a new CAT, the shop might scan the vehicle and graph the B1S1 and B1S2 O2Ss. If the rear (B1S2) has significant activity,

the bad CAT would have been verified. Figure 16–2 shows a graph representation of the two O2Ss. This vehicle has the check engine light on and has a stored P0420 CAT code.

B1S1 (top graph) has lots of activity, as it should. Remember B1S1 is the front O2S for bank 1 and is the oxygen sensor in charge fuel control. But look at the bottom trace. It is the graph for the B1S2 O2S, the oxygen sensor that is after the CAT. It should be a flat line that may trend slightly up or down. In our vehicle it has activity that is more representative of a front O2S. The technician now has two indicators that the CAT will be functioning after repair. He can look for a flat B1S2 graph prior to taking the vehicle out to reset the monitors. The baseline for this good repair included graphing the rear O2S. Once the CAT is functioning correctly, it will be flat. The ultimate authority that the repair is complete is still the monitor, but the ability to test within the shop is an important time and money saver. Once the technician is confident that the repair

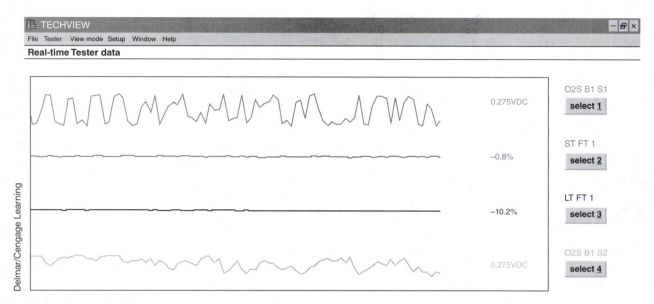

Figure 16-2 Many scanners will graph sensor values over time.

is complete, setting the monitors becomes the final piece of the repair.

Another CAT example is taken from a vehicle that has dual CATs. The car came in with one bad CAT (bank 2). The graph shown in Figure 16–3 was taken after the repair and replacement of the CAT.

Notice that the B2S2 O2S is a flat line. If you compare it to the B1 graphs, they are the same. This vehicle is repaired, and the graph function as a baseline proves it. All that remains is to run the monitors to completion and not generate any additional DTCs, and the vehicle is repaired.

The advantage of graphing data is that it gives a graphic representation of data over a period of time. Any data that changes can be graphed.

USING FUEL TRIM AS A BASELINE

The use of fuel trim as a baseline is not new to the automotive field. Ever since oxygen sensor signals have been used on vehicles, their signal has been used to adjust the delivery of fuel. This adjustment is called fuel trim and will take two forms; short term and long term. Short-term fuel trim (STFT) is the series of adjustments that the PCM will use to get the signal from the oxygen sensor back in control. If the system appears to be running rich, the PCM will cut back on the injection pulse width until the O2S averages around 450 mV. The short-term fuel trim adjustments will take place quickly so the system can respond to changes as

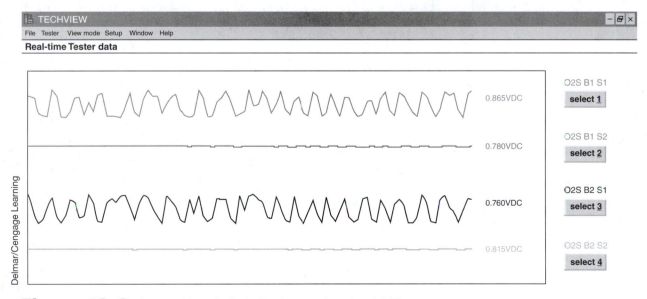

Figure 16-3 Graph of four O2Ss indicating two functional CATs.

soon as possible. These changes are not saved in any PCM memory. Once the PCM has adjusted short-term fuel trim and the vehicle is back in fuel control, the trim values will be transferred over to long-term fuel trim (LTFT) where it will be saved.

The only difference between STFT and LTFT is the time frame. STFT is used for immediate response to problems and is not stored, whereas LTFT is a result of transferring the STFT values over to a memory location that will be saved. There are hundreds of calculations done for various load and speed cells that the engine operates in. Each cell can have its own calculation, which allows for maximum control. For example, a vehicle has a simplified fuel trim consisting of 1000, 2000, 3000, and 4000 RPM and load variables of 15, 10, 5, and 0 inches of vacuum (for load calculations). Each speed has four load possibilities. So with four loads and four different speeds available, there are 16 different fuel trim calculations. Each cell of short-term fuel trim

can be different because each cell represents a different set of speed/load calculations. Some scanners display the cells in chart form, as shown in Figure 16–4.

In this example, there appears to be a problem that shows up at 2000 and 3000 rpm and engine loads of 5 in and especially at 0 in vac load (wide open throttle). All numbers are positive, so the PCM is adding fuel to compensate for a lean problem. Remember that fuel trim can be a very good indicator of either something wrong or something repaired. In this way, it can be used as a baseline.

Let's look at a simple example. A vehicle arrived at a local shop with the check engine light on and a customer complaint of rough idle and lack of power. The technician felt the 4-cylinder engine was running on 3 cylinders. In theory, this should cause a P030X DTC, indicating a misfire, but in this case there was a P0201 DTC pointing to a cylinder #1 fuel injector circuit problem. To determine a baseline, the technician noted both the check engine light and fuel trim numbers shown in Figure 16–5.

At 823 rpm, STFT was at 0.0% but LTFT was at 14.8%, indicating that lots of additional fuel is being added to the mixture and that the problem has been around for a while. The trim values started with around 15% STFT to get back into fuel control. Once the PCM was able to get the engine back into fuel control, it transferred the trim from short term to long term, where it currently sits at +14.8%.

	15 in vac	10 in vac	5 in vac	0 in vac
1000 rpm	2%	4%	8%	11%
2000 rpm	3%	5%	10%	12%
3000 rpm	3%	5%	10%	12%
4000 rpm	2%	5%	8%	8%

Figure 16–4 Fuel trim changes are based on load and engine speed. Delmar/Cengage Learning

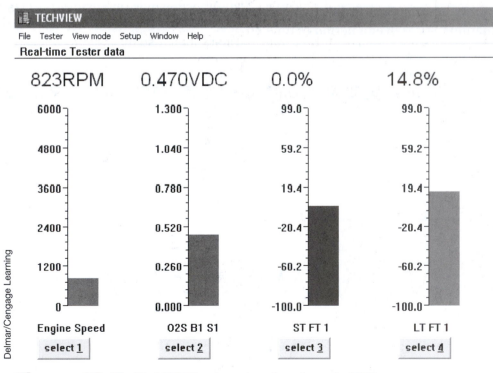

Figure 16–5 All of STFT has been transferred over to LTFT.

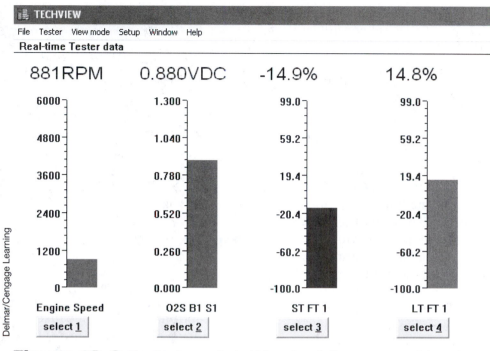

Figure 16-6 Equal but opposite fuel trim values indicate a successful repair.

The technician diagnosis was a bad fuel injector circuit from a partially melted connector on the fuel injector for cylinder #1. After he repairing the circuit with a new connector, the fuel trim values were again recorded and are shown in Figure 16-6. Notice the change in short-term fuel trim?

Notice that the STFT is now a minus value (–) and just about the same value as the LTFT. With one being +14.8% and the other being –14.9%, we know that the vehicle has been successful repaired. As soon as the fuel injector begins to inject fuel, the total quantity of fuel is now too great. Remember that the PCM had adjusted the fuel delivery system with one injector not firing by adding an additional 14.8%. Once the injector starts to function, there is now too much fuel by 14.8%, so the PCM adjusts short-term fuel trim to compensate. +14.8% (LTFT) and –14.9% (STFT) equals basically 0%. The + and the – cancel out each other out. This vehicle is fixed. The technician takes the vehicle out for a test drive and to reset the monitor by driving the freeze frame, and in a short time the check engine light goes out. The MIL is the second baseline indicator that the vehicle has been repaired. Fuel trim is an excellent source of baseline information both before and after a diagnosis and repair.

BI-DIRECTIONAL CONTROL

Bi-directional control is a function that is not supported by all manufacturers. It is the ability of the technician to control something through the scanner and have the PCM actually do it. Figure 16-7 is a simple but useful example of bi-directional control. The technician has sent to the PCM a signal to have 1100 RPM.

We know that bi-directional control is in use because of the 1100 located in the lower right corner of the PID box. This is the signal from the scanner to the PCM. Once the PCM recognizes the 1100 as an input, it will respond with the "Desired Idle 1100 RPM" that is the second PID displayed. The PCM will now adjust the idle air control until 1103 (the top PID). In this example the technician directed the PCM to raise the idle speed through the scanner. The PCM was able

Figure 16-7 Bi-directional control requests an idle of 1100 rpm and results in the PCM responding with 1103 rpm.

to actually change the idle speed. Think of all the information that this bi-directional control gave us. Through it, we know the PCM is capable of adjusting idle speed, the wires and connections from the PCM to the IAC (idle air control) valve are good, and the IAC can actually change position to adjust idle speed. All of this diagnostic information was achieved without leaving the driver's seat through our connection with the DLC.

A second example involves a vehicle that arrives at repair facility with a low EGR flow DTC (P0401). We know the EGR monitor is functioning, and after scanning for monitor status, which is a part of every diagnosis and repair, the technician uses bi-directional control to see if he can cause the EGR system to function. Figure 16–8 shows the before PIDs.

The technician has chosen to use engine vacuum as his indicator of EGR flow. If through the use of bi-directional control he is able to open the EGR and have flow, the engine vacuum will drop (Figure 16–9).

Notice the "FLW" in the lower right corner of the PID box. It indicates that the scanner has requested EGR flow. Each push of the up arrow key on the scanner will cause an additional 10% EGR flow. As the technician requests EGR flow, the engine vacuum PID does not change. This indicates that the system is not functioning, and confirms the DTC. After repair of the system, the technician can again use bi-directional and see if his repair has actually fixed the problem. Bi-directional control will allow him to control the EGR from the seat of the vehicle without having to run the monitor. Once he is sure that he has flow, the DTC can be cleared and the monitors run to complete the diagnosis and repair.

Bi-directional control is becoming more common as model year changes occur. It is a good idea

Figure 16-9 A vacuum reading with EGR flow.

to identify if the vehicle you are working on has bi-directional control because it can save a large amount of diagnosis time. The final baseline is always the monitors, but it takes precious time to get them to run. Vehicles with bi-directional control have the advantage that they can be accurately tested in the service bay prior to hitting the road for monitor setting. There is no greater waste of diagnosis time and money than to take a vehicle out that you think is repaired and spend an hour getting the monitors to run, only to have the MIL come back on and the same DTC set. The use of bi-directional control can be a great time-saver because it allows you to energize various systems with your scanner without leaving the shop.

OTHER BASELINE INDICATORS

There are additional indicators that the vehicle is fixed. Virtually anything that indicates a problem can be a baseline: scan data that shows a problem, using a DSO on various sensors or systems, or even how the technician "feels" the engine is running. Just remember that the ultimate baseline indicator is the monitoring system. The goal of any OBD II repair must be to satisfy the monitor. If the monitor runs and the check engine light remains off, the vehicle is repaired.

Throughout this text we have emphasized maintaining control of the vehicle, especially where an emission failure has occurred. It is not recommended to allow the customer to set the monitors. Frequently the enabling criteria are so complicated that it is virtually impossible for the monitor to run under "normal" driving. Setting the monitors must become part of every repair. If the monitors have run and the MIL

Figure 16-8 A vacuum reading with no EGR.

remains off, the vehicle can be returned to the customer with the knowledge that it is repaired. The baseline of the monitor has shown that it is repaired. If, on the other hand, the vehicle is returned to the customer, allowing her to set the monitors and two days later the last monitor runs and sets a DTC, she has to return the vehicle to the shop with the message that it is not fixed. This undermines customer relations. Always run the monitors to completion as part of the total diagnosis and repair job.

CONCLUSION

In this chapter, we showed various baseline techniques that can indicate when a vehicle system has been repaired successfully. We examined the use of fuel trim, the graph data function that some scanners have, and the use of bi-directional control as a baseline technique. We also discussed the use of the monitoring system as the ultimate baseline technique and emphasized that no repair job is complete until the monitors have run to completion and the check engine light remains off with no additional DTCs stored.

REVIEW QUESTIONS

1. Two technicians are discussing baseline techniques. Technician A states that a baseline will show when the repair is complete. Technician B states running the monitors is an acceptable baseline. Who is correct?

   a. Technician A only

   b. Technician B only

   c. Both Technician A and B

   d. Neither Technician A nor B

2. Short-term fuel trim goes to –22% after a repair. Long-term fuel trim is at +21%. This means

   a. the vehicle has been successfully repaired

   b. it is time to run the monitors

   c. it is time to clear all memories

   d. the vehicle still has a fuel issue

3. Technician A states that short-term fuel trim is designed to get the system back in fuel control quickly. Technician B states that long-term fuel trim values will be stored in keep-alive memory. Who is correct?

   a. Technician A only

   b. Technician B only

   c. Both Technician A and B

   d. Neither Technician A nor B

4. Using the monitor system to baseline the vehicle involves

   a. clearing memories and following the enabling criteria

   b. using the monitors to generate DTCs that still remain

   c. after the repair is complete, running the monitors to completion to see if additional DTCs are present

   d. all of the above

5. Technician A states that long-term fuel trim will slowly be transferred to short-term fuel trim as the enabling criteria are met. Technician B states that long-term fuel trim will begin to change once short-term fuel trim is stabilized. Who is correct?

   a. Technician A only

   b. Technician B only

   c. Both Technician A and B

   d. Neither Technician A nor B

6. Bi-directional control is

   a. following the enabling criteria

   b. using the scanner to control something

   c. a PID that will show the result of a monitor running

   d. used to display fuel trim

7. The graph data function is being used. Technician A states that this allows the small changes over time to be graphically represented. Technician B states that only manufacturer's data PIDs can be graphed. Who is correct?

   a. Technician A only

   b. Technician B only

   c. Both Technician A and B

   d. Neither Technician A nor B

8. The enabling criteria are followed and a DTC is set. This is after a repair. This indicates

   a. the vehicle is repaired

   b. the enabling criteria were not followed closely

   c. the vehicle is not repaired completely

   d. the vehicle will probably be rejected when it is brought for an emission test

9. A scanner is used to open the purge solenoid and close the vent solenoid. This is an example of
   a. local area networking
   b. a monitor being run
   c. a non-continuous monitor
   d. bi-directional control

10. Two technicians are discussing fuel trim. Technician A states that fuel trim values will be based on engine load. Technician B states that fuel trim values will be based on engine speed. Who is correct?
    a. Technician A only
    b. Technician B only
    c. Both Technician A and B
    d. Neither Technician A nor B

CAN AND OBD II

OBJECTIVES

At the conclusion of this chapter you should be able to: ■ Identify from a wiring diagram a vehicle that is CAN equipped ■ Recognize the different types of CAN ■ Diagnose, using an ohmmeter, a CAN system with termination resistors ■ Identify, from a wiring diagram, the gateway and explain its purpose ■ Use a DSO on the signal from a single-wire CAN ■ Use a DSO on a two-wire CAN ■ Request a module roll call by using a scanner on different CAN systems ■ Identify which module is either not communicating or generating DTC(s) ■ Explain why certain modules are on different systems ■ Use the CAN vehicle interface (CANDI module) ■ Use the breakout box for scope and DMM readings

INTRODUCTION

Controller area network or CAN has been a welcome addition to OBD II. It has brought some additional capabilities and a few difficulties to the diagnosis and repair of OBD II vehicles. CAN is a communications protocol and takes many different forms. In this chapter we will address some common traits to most systems and then examine one system in detail. It is important to note that a CAN-equipped vehicle still has monitors and requires enabling criteria to set them. Additionally, a CAN-equipped vehicle will still generate DTCs and capture freeze frame, so much of your understanding of OBD II will not change. CAN will add internal communication within the vehicle that, should it fail, needs to be diagnosed and repaired. CAN is a very complex system with many variables from manufacturer to manufacturer. It is not the intent of this chapter to completely cover CAN. Entire books can be written on the subject. Instead, it is the intent to include CAN as an integral piece of OBD II and look at its effect on the diagnosis and repair of CAN-equipped OBD II vehicles.

WHAT IS CAN AND WHY IS IT BEING USED?

CAN is a communications protocol that allows communication between both individual modules and, through the DLC, external diagnostic equipment. If your scanner is CAN capable, it will be able to communicate in much the same manner as it does with a non-CAN–equipped system. Figure 17–1 shows a typical vehicle with multiple modules.

This illustration indicates that this vehicle has 14 individual modules or control units, each capable of performing some limited diagnostics. To accurately test each module, if this vehicle were not CAN equipped, would require the technician to isolate the module, test for power and ground, and do some type of functionality test. This obviously takes time—time that might be better spent if the individual modules could communicate to a central place and "talk" to a scanner. The basic function of CAN is to link together the modules in a fashion that allows the modules, called controllers, to communicate along a common circuit. In this way the technician can connect to the DLC and receive information about each controller. The savings in time is considerable.

Think of CAN as a group of individuals within earshot of one another, as shown in Figure 17–2. Each person can send out a status report by talking loudly. In this way, everyone who needs to know the status can "hear" the communication and add to it. Notice that our person #1 has a phone. He can talk externally to someone and relay the information from the individuals in the room. CAN is similar in that all modules can communicate status to a central module, frequently the dashboard or the body control module

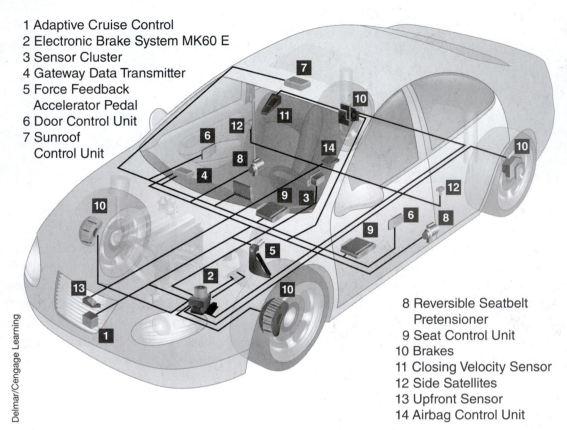

1 Adaptive Cruise Control
2 Electronic Brake System MK60 E
3 Sensor Cluster
4 Gateway Data Transmitter
5 Force Feedback
  Accelerator Pedal
6 Door Control Unit
7 Sunroof
  Control Unit

8 Reversible Seatbelt
  Pretensioner
9 Seat Control Unit
10 Brakes
11 Closing Velocity Sensor
12 Side Satellites
13 Upfront Sensor
14 Airbag Control Unit

Delmar/Cengage Learning

Figure 17-1 A vehicle with 14 modules linked together through CAN.

#1

Delmar/Cengage Learning

Figure 17-2 These five people can all talk and listen to each other. The gateway (#1) can communicate outside the group.

(BCM), but only one module can "talk" on the phone (a scanner) You may see the word *node* rather than module. A node or module is essentially the same thing with a different name. Frequently the talking node or module is referred to as the *gateway*. It is the gateway to the scanner or other communication devices, so it is named for its function.

TOOL REQUIREMENTS

CAN does have at least one tool requirement, and that is a scanner that will function within the system. A straight OBD II–only scanner will not function outside of the generic connection. Most of the better aftermarket scanners can connect to and "listen" in on the CAN communication bus if they are up-to-date with software that is CAN compliant. Some scanners require the addition of a CAN vehicle interface like that shown in Figure 17–3. These modules

Controller Area Network
CAN/OBD II
Vehicle Interface

Vetronix   P/N 02002211

Delmar/Cengage Learning

Figure 17-3 An interface may be required to communicate with the bus.

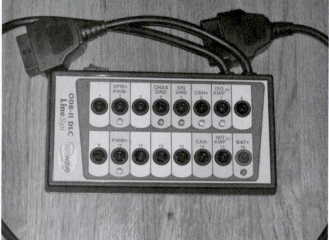

Figure 17-4 A DLC breakout box is helpful to capture patterns from the bus.

are frequently referred to as CANDI (CAN Diagnostic Interface) module. This allows the scanner to communicate with the various modules or nodes that are on the bus. The bus is the name for the communication wires that connect the various modules together.

Another handy tool to have available for CAN operation is a DLC breakout box like that shown in Figure 17–4.

The breakout box allows the connection of a scanner and at the same time the ability to connect to individual pins of the DLC. It may even have status lights like the one shown. If there are power and ground available at the DLC, the red LED at pin 16 lights as do the two green LEDs on pins 4 and 5 (the DLC grounds). The other lights on the box will flash if there is communication on the various CAN lines or bus. The DLC has very sensitive pins, especially to probing. If the pin is elongated by testing for voltage or a signal, it will likely prevent the scanner from communicating in future testing. The breakout box allows connection indirectly to the pins. In this chapter any voltmeter reading or DSO pattern was obtained at the DLC breakout box at the same time that a scanner was connected.

TYPES OF COMMUNICATION PROTOCOL

CAN communication takes place over either a one- or a two-wire system. The communication is called *serial data,* and the circuit it communicates over is called the *bus.* The speed at which the bus communicates is the class rating of the system. A class "A" bus is a slow communication (low speed) and is generally used for simple functions where speed is not a requirement. Class A communications might include the nodes for power windows, power seats, door locks, etc. A class "B" bus will operate much faster and is generally relied on to carry more complex information at faster speeds. Class B might include transmission control, the instrument panel, and maybe air conditioning/heating functions. This leaves only the powertrain control module, airbags, and perhaps anti-lock brakes where speed is very important and necessary. These communications will be handled by class "C," which is found on most vehicles after 2008. The future will no doubt bring even faster bus speeds. It is common to have more than one bus on a single vehicle. Frequently you will find a low-speed and a high-speed system, each doing different functions based on the speed of the modules communicating.

You will frequently find a CAN system referred to as a local area network or LAN. For instance, GM is currently using both a low-speed LAN and a high-speed LAN.

Figure 17–5 shows a 2007 Chevrolet Corvette's high-speed communication bus.

Notice that this is a two-wire system usually referred to as a *twisted pair,* based on the fact that the lines that surround the communication bus indicate that these two wires are twisted. At the top left of the diagram is the DLC, with pins 6 and 14 represented. Just under the DLC the circuits split into parallel circuits with one going to the vehicle communications interface module (VCIM). Toward the bottom of the diagram below the VCIM there is a 120-ohm resistor that you need to keep in mind. The other parallel circuit coming off of pins 6 and 14 of the DLC goes first to the body control module (BCM then to the electronic brake control module (EBCM), then to the transmission control module (TCM), and finally to the engine control module (ECM). Notice that within the ECM there is another 120-ohm resistor.

Figure 17–6 is another example of the use of two 120-ohm resistors wired in parallel. The first one is in the BCM off of pins 6 and 14.

The other 120-ohm resistor is located within the PCM. Ohm's law states that two equal resistors wired in parallel will result in a total resistance of one-half of the individual resistance. In other words, two 120-ohm resistors wired in parallel results in a total resistance of 60 ohms. You will see the importance of this in the future. These resistors are called *termination resistors.* They are primarily

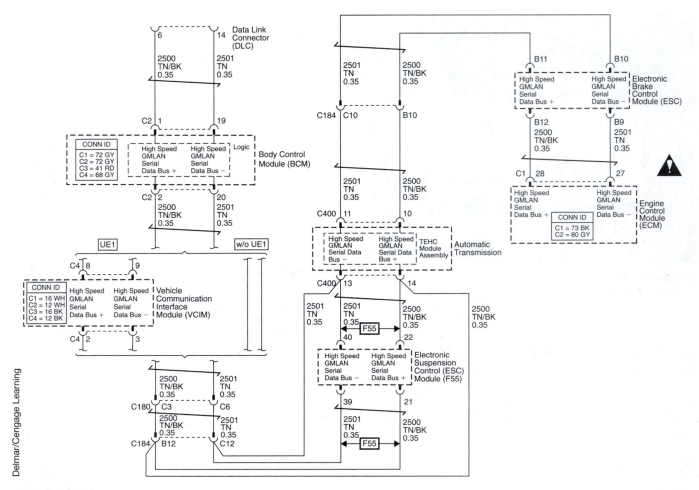

Figure 17-5 A 2007 Chevrolet Corvette uses a high-speed bus.

along as a check of the ability of the system to communicate. Sometimes they are internal, within a module, but sometimes they are found within a fuse box.

Figure 17–7 shows the 120-ohm resistor located within the rear fuse block on a 2008 Buick. A simple but important check of a system that has termination resistors is to remove power by disconnecting the battery, connecting an ohmmeter between the terminals 6 and 14 in the DLC breakout box, and measure the total resistance.

Remember that two 120-ohm resistors wired in parallel will result in a total resistance of 60 ohms. Figure 17–8 shows the results of connecting between pins 6 and 14 of the DLC (battery disconnected).

61.4 ohms indicates that both termination resistors are connected and there is a complete circuit available for the communication bus. This simple but very effective test should be done on a vehicle that has the termination resistors wired in parallel that is

generating a communication DTC. Remember earlier in this text we discussed P0 DTCs and identified them as generic. We also identified P1 DTCs as manufacturer generated. Once we get into CAN or LAN systems the "U" DTC appears, indicating a communications problem. What do you think is wrong if the ohmmeter reads 120 ohms? Apparently one of the termination resistors is not in the circuit, and we are measuring only one. An infinity reading would indicate an open circuit, and zero ohms would indicate a short circuit on the bus. Much of our diagnostics today does not utilize the DMM (digital multimeter), and we see increased use of scanners and DSOs. However, this is a simple check of the capability of the bus to communicate. It does not mean it *will* communicate, simply that it has the ability. Also remember that the termination resistors can be anywhere in the system. They can be in a module, within the wiring harness, or in a fuse box. Always refer to the wiring diagram for their location.

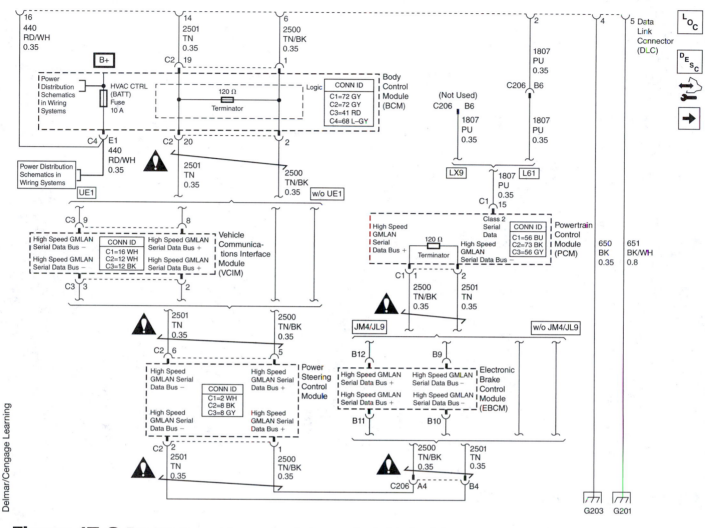

Figure 17-6 Two 120-ohm resistors are located across the twisted-pair bus.

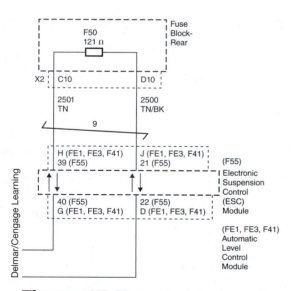

Figure 17-7 Sometimes the termination resistors are located in a fuse block.

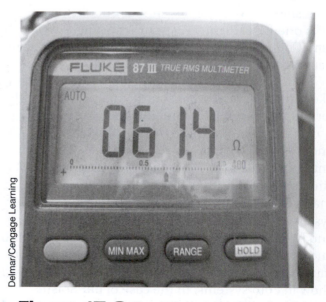

Figure 17-8 Two 120-ohm resistors wired in parallel results in a total resistance of 60 ohm (battery disconnected).

Delmar/Cengage Learning

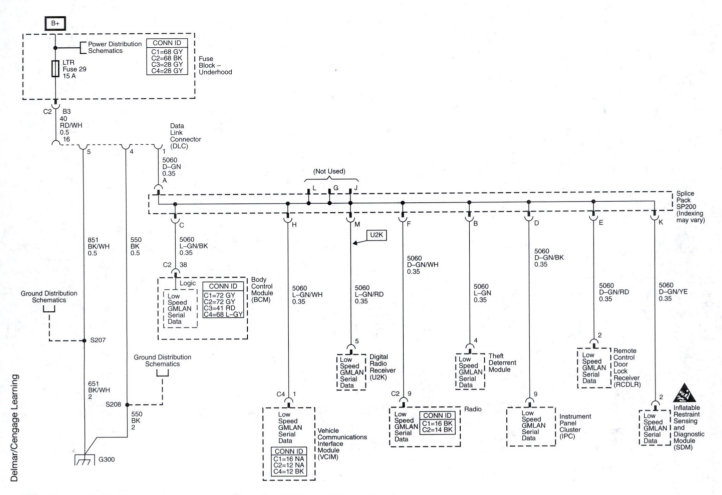

Figure 17-9 A single-wire system with eight modules on the bus.

A single-wire system is similar but obviously only uses one wire. Figure 17–9 shows a single-wire system. Communication between modules and a scanner is accomplished at pin 1 of the DLC.

Notice also that the modules include slower items like the radio, the remote control door locks, theft deterrent module (TDM), and inflatable restraint diagnostic module. Compare this to a high-speed LAN, and it is obvious that the list of devices will include anti-lock brakes, the PCM, the transmission, etc. This single-wire diagram is from a 2008 Pontiac Solstice that will have both high-speed and low-speed LAN systems onboard. Low speed will be available at pin 1, and high speed (two wire) will be available on pins 6 and 14. The two systems will also be capable of communicating with each other because one module will show in both systems. In the case of our Solstice, the BCM is part of both the high-speed and the low-speed LAN. Figure 17–10 shows the high-speed bus diagram

and is used to indicate what type of system is in use, where the termination resistors are, which terminals are communication terminals, and what modules are on the network.

This high-speed LAN has a BCM, the VCIM, the power steering module (PSM), the EBCM, and the PCM all on the bus communicating on pins 6 and 14.

COMMUNICATION VIA THE BUS

A DSO is a good device to actually look at communication as it occurs. For instance, if we connect a DSO between pins 6 and ground of a two-wire high-speed LAN, we should see Figure 17–11.

The pulse runs from 2.5 to 3.5 volts on pin 6 of the DLC. Figure 17–12 shows the DSO pattern on pin 14 and ground. Notice that this pulse runs from 1.5 to 2.5 volts.

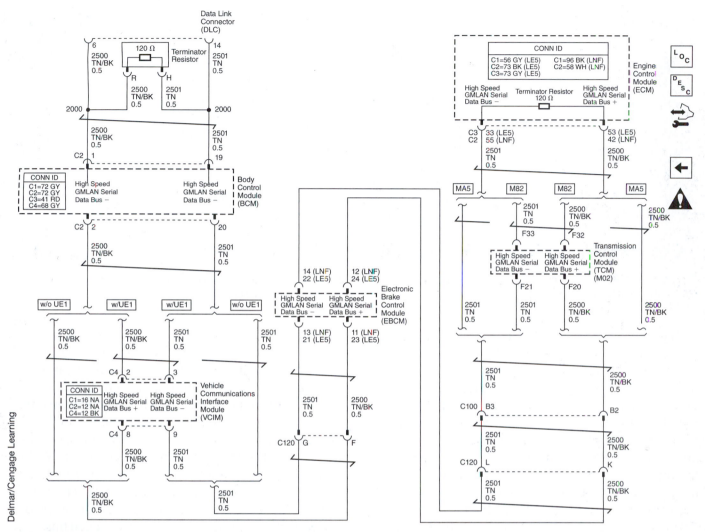

Figure 17-10 A high-speed bus communicating on pins 6 and 14 of the DLC.

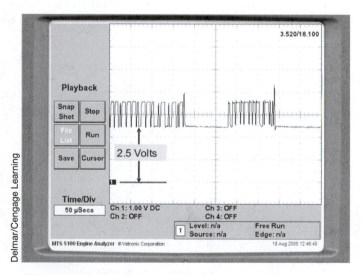

Figure 17-11 Pins 6 and ground on a high-speed bus. Runs from 2.5 to 3.5 volts.

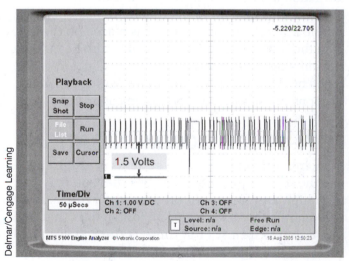

Figure 17-12 Pin 14 and ground on a high-speed bus runs from 1.5 to 2.5 volts.

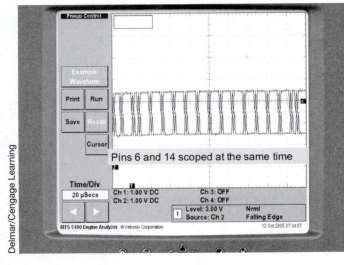

Figure 17-13 Pins 6 and 14 dual traced reveal a mirror image of data.

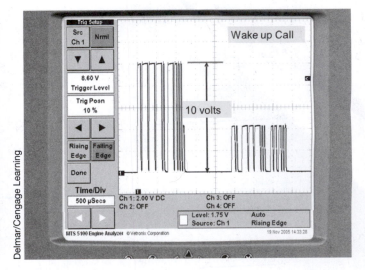

Figure 17-14 A single-wire system with a 10 V wake-up call and data.

If we dual-trace pins 6 and 14, we see Figure 17–13. Notice that the patterns are mirror images of one another. Each data bus is transmitting the same information but in opposite voltage transitions. These voltage changes both in the positive and negative direction are pulses that can be interpreted by the gateway, any modules that need the information and/or a scanner. This is the communication along the bus. One of the advantages of a two-wire mirror image data bus is that if one leg were to have a fault, the other leg can still deliver all the information. There are two parallel paths each sending the same data along the bus. Because there are five modules or more on the bus, each module "hears" the data being sent along the bus and decides whether it needs the information or just sends it along. Think about the modules and what data they might share. Vehicle speed is a simple example. If vehicle speed is placed on the bus, then the BCM, the ECM/PCM, and the brake module can all "share" the information. They all need vehicle speed but for different reasons. The sharing of data is a major responsibility of the communication bus.

LOW-SPEED COMMUNICATION

Remember that not all CAN systems are high speed like the one we just looked at and that you may find both high and low speed on the same vehicle. Pin 1 will again show a series of on/off signals that will come in two forms. First there is the wake-up call from the bus. Figure 17–14 shows this.

Notice the 10-volt fast on/off signal. This is the wake-up call to the modules indicating that they should begin sending both status and function information. Once the module begins talking, the system will typically be at a lower voltage. Our GM example shows some activity

after the wake-up 10-volt activity that is around 4.2 volts. This indicates that the system is indeed "talking" on the information bus. You can see that this is a slower form of information transfer by looking at the speed of the DSO/DIV. Our high-speed example was captured at 20 microseconds per division (20 millionths of a second per division), while our low-speed example was captured at 500 microseconds per division. The low speed is actually about 25 times slower than the high speed. Don't forget that the manufacturer decides which modules to communicate on the high-speed bus and which to communicate on the low-speed bus.

It is possible to look at both high speed and low speed on the same vehicle using a three-channel DSO, although the different speeds may make it difficult to see the voltage transitions. For this example, let's use a 2005 Cadillac CTS. Figure 17–15 shows the results

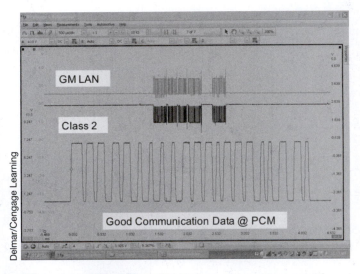

Figure 17-15 A GM high-speed and low-speed system at 1, 6, and 14 pins of the DLC.

of connecting a DSO to pins 6, 14 (high-speed LAN), and 1. Notice that the top signal goes up from a baseline. The second or middle signal goes down from a baseline, and the third or bottom signal is the familiar pulsing class 2 running from 0 to about 4.2 V after wake-up. It appears that this vehicle is communicating along the various networks available. We should have no difficulty communicating with the various gateways being used.

DTCS FOR COMMUNICATION

The familiar P0 and P1 DTCs are still present in a CAN- or LAN-equipped vehicle. However, a new set of "U" codes may be generated. The U code is for communication failures and is only available on communication bus–equipped vehicles. Remember that "U" codes are for communication problems. Either a single module or node is not communicating because it can't, or the bus will not transmit the data. Don't forget that you will need an additional communication interface or CANDI module to be able to use most scanners.

When you connect a scanner, it will first display module status by requesting a roll call on the LAN. Let's follow through with the module status request on a 2006 Pontiac G6. Initially, when you select Module Status on the start-up screen, the scanner display will ask you what you "think" is on the vehicle (Figure 17–16).

This G6 has class 2. When we select #1, the scanner will look for modules that will report in and Figure 17–17 will show on the screen.

After the scanner has established communication with the bus, it will take roll call and request

Figure 17-17 The scanner requests a roll call of modules and their status.

Figure 17-18 Five modules report in, with one generating DTCs.

any current problems from each module that will show up as DTCs. The screen will now show which modules are present and what DTCs are associated with each (Figure 17–18).

The scanner shows that there are five modules on this vehicle. Each shows with a hexadecimal number and its abbreviation and whether it has any DTCs. In our example one of the five modules present shows DTCs. If we highlight the radio, we would see that it has had communication problems in the past. You need to recognize that all five modules are reporting in with only one experiencing a problem. The modules that are able to communicate along the bus are the BCM; the instrument panel cluster (IPC); the supplemental inflatable restraint (SIR), or airbag; the theft

Figure 17-16 A scanner displays possible systems on a vehicle.

deterrent module (TDM), and the radio. Additionally, the scanner is telling us that it can communicate with the bus. This is one of the most important things to do when first diagnosing a CAN-equipped OBD II vehicle. Most systems have very good capability to perform self-diagnosis and set DTCs for each module. These DTCs are in addition to the standard OBD II DTCs that are set by running the monitors. They are, however, dependent on the network to deliver the DTC to the scanner through the gateway. In our example above in Figure 17-18, the five modules are able to communicate any difficulties they are having. They will transfer the information via the communication bus or the LAN because they answered "here" to the roll call sent out by the gateway. Their status is known because they are talking and listening on the bus.

If a DTC is present, the status changes from No to Yes and highlighting the yes lists the code and possibly the description. For this G6 there are a series of "U" codes listed in Figure 17-19.

Individual descriptions give much more detail. You will sometimes have to refer to the manufacturer's literature to decipher the DTC. Figure 17-20 shows some definitions of some common communication DTCs.

U Code Charts

Information
▸ U0073
▸ U0100
▸ U0101
▸ U0107
▸ U0121
▸ U0140
▸ U1000
▸ U1016
▸ U1024
▸ U1048
▸ U1064
▸ U1088
▸ U1096
▸ U1151
▸ U1300
▸ U1301
▸ U1305
▸ U1500
▸ U2100
▸ U2102
▸ U2103
▸ U2104
▸ U2105
▸ U2106
▸ U2107
▸ U2108

Figure 17-19 U codes indicate communication problems.

DTC DESCRIPTORS

This diagnostic procedure supports the following DTCs:

• DTC U0100 Lost Communication With Engine Control Module (ECM)
• DTC U0101 Lost Communication With Transmission Control Module (TCM)
• DTC U0121 Lost Communication With Electronic Brake Control Module (EBCM)
• DTC U0140 Lost Communication With Body Control Module (BCM)

Figure 17-20 U code descriptions are detailed explanations of communication issues.
Delmar/Cengage Learning

You can see from this abbreviated list that U codes are primarily communication codes. If a module or node cannot connect with the bus or network, it has lost communication and a U code will be set. Most U codes will turn on the MIL and cause a failure of an emission test even though they might not cause an increase in emissions. If the network cannot communicate with the BCM (U0140), then it does not know if it is having problems and therefore will turn on the MIL. With the MIL on, one of the first things a tech should do is scan for module status, paying attention to those modules that are generating DTCs and also those that have lost communication. Frequently problems within the network will cause no starts, MIL on, or drivability problems that need to be addressed. Gone are the days when the MIL being on was strictly an emission problem. Remember that the bus needs to have the ability to communicate, and the individual modules need to both listen and talk. Listen for information on the bus that is needed and then add their information and status to the bus.

Let's take a simple OBD II problem and add CAN to it. A vehicle is having fuel pump problems, and fuel trim is trying to compensate for the problem. Prior to CAN this would have been information that the PCM would use to turn on the MIL and capture freeze frame. The PCM is still involved but now might broadcast the information along the bus to other modules that might need it. Perhaps the VCIM and the BCM would receive the information because they are in charge of the MIL. When you plug your scanner into the vehicle, you would communicate with the gateway not directly with the PCM as you did in the mid-nineties. The gateway would deliver the information to the DLC along the bus. In our simple fuel pump example, the vehicle needs to have a functioning bus, VCM, PCM, and BCM to deliver the information you need regarding the MIL status.

The first check of any CAN system must be to see if the network is up and functioning. If the network is damaged, open, or shorted, then information will not be deliverable or readable by the scanner connected into the DLC. Remember that the network is the key to your diagnosis and repair of the rest of the OBD II system. Without it functioning you will not be able to communicate with the vehicle.

Additionally, in states that have emission testing, no communication will either be a reject or a fail. Both will require diagnosis and repair prior to bringing the vehicle back for a retest.

Let's connect our scanner to a high-speed bus on the 2008 Pontiac Solstice that we looked at earlier. If the network is functioning, we should get the familiar Select Network screen. In this case let's look at dual-wire CAN (Figure 17–21).

The dual-wire CAN system is high speed and involves five modules. Our scanner sends out the request for roll call, and the network sends back module status for those modules on the bus. Figure 17–22 shows the results.

This network is communicating, and we can see all three modules are present with no DTCs. If the network was down for any reason, the scanner would not be able to establish communication. In the example of the Solstice, the three modules that are high speed, the body control module (BCM), the powertrain control module (PCM), and the transmission control module (TCM) are all communicating on the network.

The Solstice is equipped with a single-wire system also, so let's go back to the roll call page of the scanner and request module status (Figure 17–23).

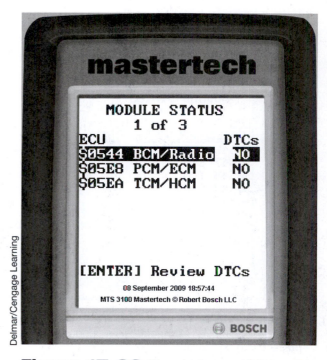

Figure 17-22 Three high-speed CAN modules report in with no DTCs present.

Figure 17-23 By selecting #3, the scanner will request a single-wire CAN roll call and module status.

Figure 17-21 Dual Wire (high-speed) CAN is selected for roll call and status.

By selecting #3, the scanner requests the same roll call and module status, but now from the single low-speed CAN side of the gateway. The results are shown in Figure 17–24.

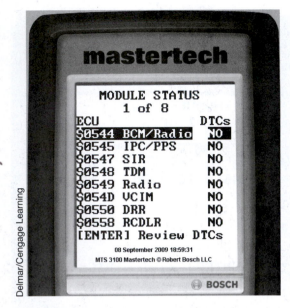

Figure 17-24 Roll call for the Pontiac Solstice single-wire CAN indicates all modules present with no DTCs.

The list shows that eight modules have reported in on the roll call with no DTCs present. Remember that this is actually three tests in one. First, we know the network is up and running, and second, we know that there are eight modules answering the roll call. The third test indicates that there are no DTCs present at this time. A comparison of this screen to the wiring diagram and all the information required to diagnose and repair a CAN-equipped vehicle is present.

Don't forget if communication cannot be established at the emission lane or in your shop, you will need to diagnose and repair the network prior to actually seeing scan data. This will add "network function" to the "repair DTC and run monitors" that were part of the repair process prior to CAN. Make sure you add the time required to the estimate to diagnose and repair the network. Customers need to understand that no communication is the first item, not necessarily the last, needing attention. For example, a CAN-equipped vehicle arrived at the emission testing center in the Midwest. The lane operator plugged into the DLC and received a "no communication" error message. He now tells the owner of the vehicle that he needs to have the error fixed and it should not be very expensive to do. The consumer takes the vehicle to a reputable shop to get the no communication fixed. The shop diagnoses an open network, and repairs it. Immediately the MIL comes on. When the shop scans for data,

they realize that the vehicle needs a new CAT—a very expensive component on this vehicle. The customer rants and raves that the emission test lane said it would not be expensive, and now the shop wants over a thousand dollars to "fix" the vehicle. If only the shop had informed the consumer that lack of communication is one problem that they could repair, but after repair and communication have been established, any additional problems requiring repair will show up.

In this true story the consumer refuses to have the vehicle CAT replaced and brings the vehicle back to the emission test lane, where it receives a fail. The unfortunate part of the problem is the shop that repaired the lack of communication was tagged with the emission failure. This dropped down their repair grade. Eventually the consumer brought the vehicle to another shop and had the CAT installed. The second shop received the plus for their repair grade because the new CAT allowed the vehicle to pass the emission test. Communication is extremely important both on the vehicle and between the consumer and repair facility.

CONCLUSION

In this chapter, we looked at some common controller area network systems beginning with class 2 and ending with a dual-wire high-speed system. We emphasized the importance of the communication bus, beginning with the gateway. We examined how to diagnose a bus that is not able to communicate with a simple digital multimeter and the wiring diagram. We looked at the low-speed and high-speed data with a DSO and a breakout box. Additionally, we looked at how the scanner displays module status and DTC associated with the network.

REVIEW QUESTIONS

1. Two technicians are discussing CAN. Technician A states that single-wire systems can be accessed through the DLC. Technician B states that high-speed CAN may be a dual-wire, twisted pair from the DLC. Who is correct?

   a. Technician A only

   b. Technician B only

   c. Both Technician A and B

   d. Neither Technician A nor B

2. A wiring diagram for a GM vehicle shows a twisted pair coming off of pins 6 and 14 of the DLC. This indicates that this is a

   a. class 2 CAN

   b. low-speed CAN

   c. high-speed CAN

   d. single-wire CAN

3. A vehicle is being diagnosed for no communication. The wiring diagram shows two termination resistors, each of 120 ohms. Technician A states it is necessary to remove power from the system by disconnecting the battery. Technician B states the total resistance of the two resistors should be 240 ohms. Who is correct?

   a. Technician A only

   b. Technician B only

   c. Both Technician A and B

   d. Neither Technician A nor B

4. The purpose of the gateway is to

   a. provide communication to a scanner

   b. store all DTCs

   c. identify what system the vehicle has

   d. be the central link to all systems on the vehicle

5. A DSO is connected to a single-wire system. If the system is communicating, the DSO should show a

   a. series of negative voltage spikes

   b. flat line at zero volts

   c. flat line at battery voltage

   d. series of on and off signals pulse width modulated

6. A DSO is connected to a two-wire (pins 6 and 14 and ground) at the DLC. Technician A states that there will be a single pulse that is pulse width modulated. Technician B states that there should be a mirror image signal on the two channels. Who is correct?

   a. Technician A only

   b. Technician B only

   c. Both Technician A and B

   d. Neither Technician A nor B

7. A scanner is calling for a roll call on a single-wire CAN system. After the roll call, the scanner should show

   a. pulsing voltage signals

   b. module status and DTCs

   c. serial data from input sensors

   d. the communication data from only modules that have active DTCs

8. Technician A states that modules that do not require frequent and fast updates will normally be in the low-speed CAN system. Technician B states that modules that require frequent and fast updates will normally be in the high-speed CAN system. Who is correct?

   a. Technician A only

   b. Technician B only

   c. Both Technician A and B

   d. Neither Technician A nor B

9. A breakout box is preferred to backprobing the DLC because

   a. DLC pins are small and fragile

   b. a scanner can be connected at the same time

   c. most data required for CAN diagnosis is present in the DLC

   d. all of the above

10. Technician A states that a CAN interface module is required to be able to use a breakout box on a DLC. Technician B states that the breakout box is required if module status is requested through a scanner. Who is correct?

    a. Technician A only

    b. Technician B only

    c. Both Technician A and B

    d. Neither Technician A nor B

USING A DSO ON AN OBD II SYSTEM

OBJECTIVES

At the conclusion of this chapter you should be able to: ■ Understand that the DSO reads voltage over a period of time ■ Recognize how to set up the DSO to see various signals ■ Test an oxygen sensor for rich voltage accuracy, lean voltage accuracy, and reaction time (speed) ■ Test a fuel pump and calculate its speed against a known standard ■ Test an engine's compression using a DSO and a current probe ■ Test an ignition system on a vehicle with DTCs ■ Test fuel injectors determine if the injector is in need of cleaning ■ Test a DLC for its ability to communicate during an emissions test

INTRODUCTION

The use of the digital storage oscilloscope (DSO) has increased since its beginning during the early 1990s. OBD I (1980–1995) did not generate many DTCs, and technicians frequently found themselves trying to look within the system. The DSO became the mainstay of the industry during the early '90s. Unfortunately, many technicians believe that within OBD II the scanner replaces the DSO. This is not necessarily the case. It is true that the first tool probably used will be your scanner; however, there will be times when the signal from a particular sensor or output will need to be analyzed beyond the capability of the scanner. The DSO is a natural tool to use.

Let's begin with a quick review of the specifics of the DSO. Figure 18–1 shows the DSO screen and arrows indicating that voltage is up and down, while time is displayed across the screen from left to right. The basics setting for the DSO involve the time per division and the voltage per division. The DSO shown has ten divisions for time running from left to right and eight divisions of voltage running from bottom to top. If, for instance, you were to set the voltage to 1 volt per division, you would have a total of 8 volts to "see." Battery voltage would not be able to be displayed with a setting of 1 volt per division. Changing the voltage per division to a greater value is required, and the change would follow the 1 – 2 – 5,

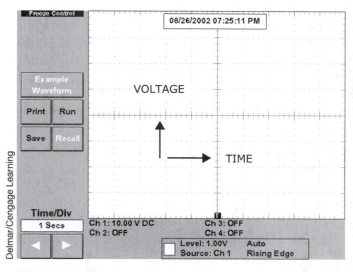

Figure 18-1 A DSO will display voltage changes over a period of time.

sequence. So if we are at 1 volt per division, the next higher value will be 2 volts per division. 2V per Div means there will be a total of 16 volts of measurement available, which will make measuring battery voltage possible. The time per division usually is set to display the entire signal of the questionable sensor. If the sensor is generating multiple signals that are cylinder oriented, like a CKP (crankshaft position sensor), the screen should show all signals. Look at the signal displayed in Figure 18-2.

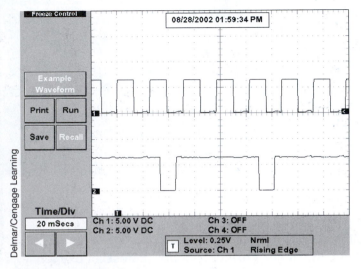

Figure 18-2 The DSO can display more than one signal at a time.

This figure shows a dual display for a CKP that generates a 3X and a 1X signal every time the crankshaft makes one rotation. The bottom (channel 2) shows two signals, indicating that the crankshaft has rotated at least twice. It is important to make sure the display shows all of the signals you are trying to analyze. If the signal generates 18 pulses per crankshaft rotation, then at least 18 signals should be displayed. When dealing with sensor readings, you will mostly use the auto set feature that most DSOs have to get a signal on the screen and then adjust both time and voltage to fill the screen and show enough signal changes.

OXYGEN SENSOR TESTING

One of the most important signals required by the OBD II system is the oxygen sensor, and although it is true that the monitoring system "should" find and identify the bad ones, it is just as true that bad O2Ss may shut down the monitoring system. This results in monitors not running and no DTCs. Remember that the only way to generate DTCs on an OBD II vehicle is to run the monitors. If the O2S is shutting down the monitoring system, then no DTCs can be generated. Frequently on V-type engines, one front O2S is bad and one is good, shutting down the monitoring system. You need to be able to identify out-of-spec O2Ss, and the DSO is the best method.

There are three factors that need to be independently tested: rich voltage accuracy, lean voltage accuracy, and speed. This is easily done by injecting a controlled amount of propane into the intake and monitoring the O2S signal.

The setup of the DSO is the first consideration. The volts per division (V/Div) should be set for 200 mV/Div with the zero volts line one or two

divisions from the bottom of the screen. Set the time for 500 mS/Div (½ second). This means that the total time on the screen will be 5 seconds (10 divisions × 500 mS = 5 seconds). When finished, the screen should look like Figure 18–3.

Next we need to tap into the O2S signal. This can be done in a variety of places, but the most logical is right at the connector for the sensor as Figure 18–4 shows.

Notice also that the ground clip is near the O2S on the engine. This single-wire sensor uses the ground circuit of the engine for its sensor ground. Many of the O2Ss in use will have a ground wire in the connector that should be used.

Next, we have to preset or set up the injection of propane into the manifold. This does not require that you disassemble half the engine, just that the propane

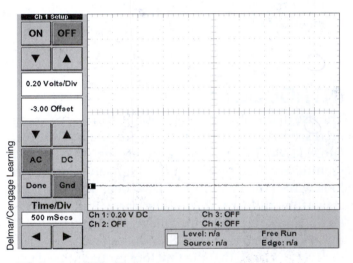

Figure 18-3 Oxygen sensor testing is best done at 200 mV/Div and 500 mS/Div.

Figure 18-4 Oxygen sensor testing requires a solid ground connection.

be injected close to, or preferably below, the throttle plates, as Figure 18–5 shows.

The propane hose goes to a fitting below the throttle plates.

The procedure is really simple. The first step is to inject enough propane to flatline the sensor voltage at 2000 engine RPM. The signal must flatline to indicate the maximum voltage the sensor can generate. If it will not flatline, increase the flow of propane. Ignore any other signal irregularities while you are injecting the propane. With the engine at 2000 rpm and the propane on, our test vehicle generates the signal shown in Figure 18–6.

With the zero line one division off the bottom and every division equaling 200 mV, the flatline part of the signal is at 875 m. Is this good or bad? If we go back to the original Bosch specification for maximum rich voltage, we find that the signal should be between 800 mV and 1 V. The rich voltage on this sensor is between the specifications and indicates that the sensor is capable of generating an accurate maximum voltage under rich conditions.

The next part of the test is to turn off the propane and watch for a flatline lean condition, as Figure 18–7 shows.

When propane is turned off, the system goes lean and the voltage drops. Again it has to flatline lean or the test is not valid. Our O2S sensor appears to be generating a signal around 50 mV, and the specification is between 0.0 V and 175 mV. Our sensor appears to be capable of generating an accurate lean voltage.

The last and perhaps the most important piece of the puzzle is how fast can the sensor go from full lean to full rich. While the signal is still lean, we will again turn on the propane. This should result in a straight upward line, as Figure 18–8 shows.

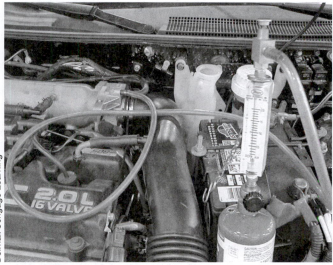

Figure 18–5 To correctly test an O2S, propane should be injected below the throttle plates.

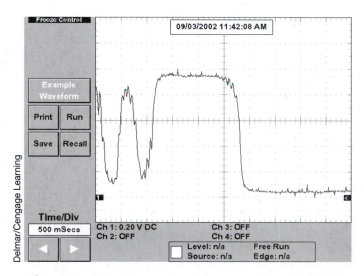

Figure 18–7 Just after turning propane off, the DSO displays a 175 mV lean voltage.

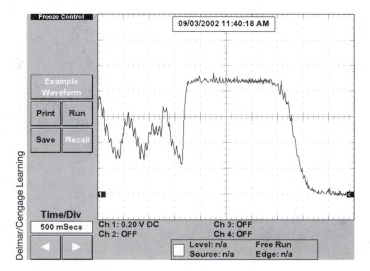

Figure 18–6 With propane flowing, the DSO shows a rich voltage of 875 mV.

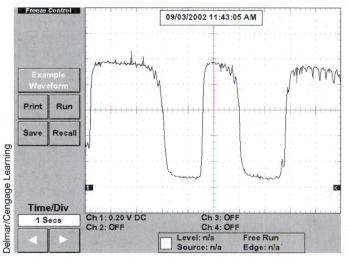

Figure 18–8 The lean to rich voltage after turning propane on indicates the speed of the O2S.

There are two transitions shown where propane was turned on: in the 5th division and in the beginning of the 8th division. Notice that the line is pretty much straight up, and with a spec of less than 100 mS, the speed of this sensor appears to be good.

If, as a technician, you make it a point to check the various oxygen sensors on a vehicle, you will find faulty ones frequently. Remember the interrelationship between the O2Ss and the monitors: many monitors require the signal from the O2Ss to be able to function. Problems within the oxygen sensors will likely result either in monitors not running to completion or in false results. It takes equality between the O2Ss on the vehicle for the monitoring system to function. Equal rich voltage, equal lean voltage, and equal speed are requirements. You will need to figure out how to test oxygen sensors, and the DSO will most likely do the job for you easily. You will not typically be able to rely on scanner data to pinpoint bad O2Ss. Sometimes mode 5 is useful as we saw in Chapter 15, but the definitive solution is to force the system rich and measure the voltage; force it lean and measure the voltage; and finally force it rich again and measure the speed. In this way you will be able to eliminate various sensors from the monitor mix.

FUEL ISSUES

The PCM on vehicles of today has linked together fuel and ignition, making it difficult to diagnose by test driving as we did years ago. A good technician needs to be able to look at ignition and fuel separately. Let's start with some fuel issues and see how the DSO can simplify your diagnosis.

Fuel delivery takes both pressure and volume. The pressure is primarily controlled by the pressure regulator. But for it to work correctly, it must have adequate volume from the fuel pump. Pressure will be controllable only if adequate volume is available. This means that lines, filters and connections must be able to flow the required minimum volume that the fuel pump is delivering. Fuel volume is a key to adequate volume. There are gauges on the market that a technician installs to indicate the volume of the pump and usually the pressure. However, a DSO can be used to calculate pump speed, and pump speed and volume are directly related. As the speed of a pump drops, so does its volume. The DSO will give us the opportunity to check the speed of the pump by using a current probe, as shown in Figure 18–9.

The current probe generates a voltage signal based on current flow. When the current probe is connected to the DSO, its output will show as a voltage on the

Figure 18-9 A low-current probe will allow the DSO to display current over a period of time.

screen. When the pump spins, it will draw DC current that the probe can measure. The pump will also generate an AC signal that will "ride" on the DC current. We want to measure this AC signal since its unique signal signature can be used to tell how fast the fuel pump armature is spinning.

With the engine running, the DSO is set to AC voltage and the current probe placed around either the positive or negative feed wire to the pump. The pattern in Figure 18–10 is the result.

The commutator inside the pump will have an even number of bars, each of which is connected to a set of windings. Each bar will generate its own signal,

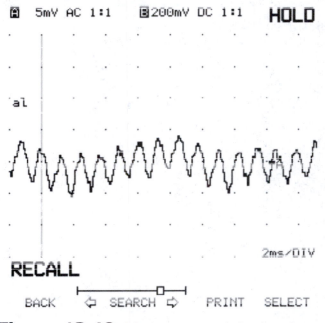

Figure 18-10 A fuel pump current signal can be used to calculate pump speed.

so every hump in the pattern represents the slight turning of the pump armature. Our test pump has an armature with eight commutator bars, which is very common. Each rotation of the pump will generate eight AC voltage waveforms. If we count off eight pulses, we have the time for one rotation. In our test pattern, the DSO has been set for 2 mS per division, and one rotation of the armature appears to take approximately 9 mS. So it takes 9 mS (.009 sec) for the pump armature to spin one rotation.

The next thing we need to do is a bit of math to determine the RPM of the pump. It is best to do this in stages. If we take the .009 sec and divide it into 1 (second), we will get the number of rotations per second. Get out the calculator and put 1 divided by .009 and you should get 111. This means that the pump armature spins 111 times in one second. Make sure you understand how we arrived at this number before you go on. Feel free to read it over until it makes sense. There are 60 seconds in a minute, so now all we have to do is multiply the 111×60 and we will get the RPM of the pump armature: $111 \times 60 = 6660$. This pump is spinning at 6660 revolutions per minute, and most manufacturers specify a minimum of 4500 rpm, so this pump is capable of delivering the required minimum volume.

Let's take another example. This vehicle was experiencing significant misfire and was generating P0300 DTCs. Fuel trims under load seemed to indicate a lean condition. Additionally, high-speed operation seemed to have the most misfire. A technician decided that fuel pump speed was a good thing to test and resulted in Figure 18–11.

The first thing to notice is the 5 mS/Div, which is two and a half times slower than the previous example (2 mS/Div). Count off the eight pulses and write down the time. It appears to be around 37 mS or .037 sec for one rotation. Even without finishing off the calculations you can see that this pump is significantly slower. 1 divided by .037 = 27 rotations per second. 27 rotations $\times$ 60 seconds = 1621. This pump is spinning at 1621 RPM, which is below the 4500 rpm minimum. Replacing the pump eliminated the misfire DTCs and the drivability problem.

Figure 18–12 shows another example of a fuel pump waveform.

This pump also has eight commutator bars in the armature. Again the DSO has been set for AC, and a current probe is measuring the current flow through the pump. The DSO is set for 2 mS/Div. Using the bottom of the trace as our indicator of each bar, count off eight pulses. How many mS of time does it take for one rotation? The measurement shows just about 16 mS or .016 sec. Again we divide the .016 into 1 to determine the rotations per second: 1/.016 = 62.5 RPS. Again we multiply the 62.5 $\times$ 60 sec to obtain the RPM: 62.5 $\times$ 60 = 3750 RPM.

So what is the status of this pump? With the general specification of 4500 RPM minimum, the pump is running too slow. What might be the problems associated with this reduced volume? The most obvious is misfire resulting P0300 DTCs; however, as the pump spins slower and volume drops, the fuel trim values will attempt to keep the system in fuel control. Many vehicle PCMs are programmed to allow only a specific amount of trim. Once the vehicle requires more trim than specified, the system locks the trim down.

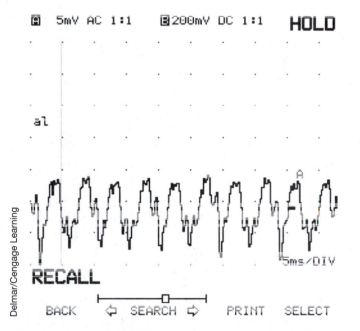

Figure 18–11 The signal from a very slow fuel pump.

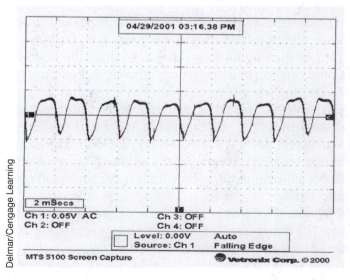

Figure 18–12 This fuel pump is spinning at 3750 RPM, which is below the 4500 RPM standard.

This allows the system to run lean and may result in DTCs and/or monitor issues. The key to understanding the system is looking at all parameters that the processor looks at and then manually looking at items that the processor does not. Fuel pump RPM is a good example of this. There is currently no vehicle that looks at pump RPM as an input. The system assumes that it is correct and that sufficient fuel volume is available for the pressure regulator to control fuel pressure for injection.

COMPRESSION TESTING

Let's look at another common use for a DSO on OBD II vehicles: compression testing. As you work on any vehicle, it is always important to keep in mind that understanding the basics of compression, ignition, and fuel is required. The modern OBD II vehicle does not have nearly the compression problems that older vehicles had, but it still is an important consideration when dealing with monitor issues or P030X problems specific to a single cylinder. An additional consideration is the fact that most OBD II vehicles have buried the spark plugs within an aluminum cylinder head. The manufacturers have stated to not take plugs out of a warm engine, yet mechanical compression testing is usually done on warm engines. So the use of a mechanical compression test is virtually impossible.

Enter the DSO compression test, which can be done on a warm engine and will give relative readings. Relative means that you will not get a specific pounds per square inch reading for each cylinder but will instead see all cylinders against one another. Additionally, you will see compression readings in amps per cylinder. To make sense of this, you have to realize that the engine, during cranking, will slow down slightly as each cylinder comes up on compression. As the engine slows down, so does the starter motor, causing it to draw more current. For compression testing you will need to have a current probe that is capable of measuring over 150 amps. Connect the probe around the starting cable, disable fuel so the vehicle will not start, use a synch probe around plug wire #1 (trigger), and crank the engine. Figure 18–13 shows a 6-cylinder engine with low compression on one cylinder.

Notice the flat spot toward the right of the screen. The "T" along the bottom edge shows where spark plug #1 fired. The upward turn of the waveform prior to the "T" is compression for cylinder #1. The firing order for this vehicle is 1, 6, 5, 4, 3, 2. If you count from the "T," you should realize that cylinder #5 is the flat

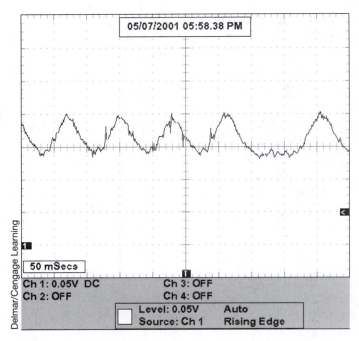

Figure 18-13 A high current probe allows compression testing. This pattern indicates one cylinder with low compression.

section. This vehicle failed the required emission test and would have stored in memory a P0305 DTC, indicating misfire in cylinder #5. This turned out to be a burned exhaust valve. After removing the head and having a valve job done on it, another compression test was performed, with Figure 18–14 being the results.

Notice that the flat spot has been replaced by pulses that are close to being equal with the rest of the pattern. A test drive, following the "drive the freeze frame," resulted in the MIL going out, indicating that the original problem had been fixed. The vehicle was retested and passed the emission test, codes were cleared (after the test), and the vehicle was returned to the happy customer.

IGNITION TESTING

Another great use of the DSO on OBD II vehicles is to look at the ignition and fuel injector patterns. Let's start with ignition and analyze the pattern for indications of ignition problems. This is not intended as a complete ignition analysis. For that, look to other texts such as *Tech One: Automotive Electricity and Electronics,* where ignition is covered more completely. For our purposes we will be concerned with the spark, including the voltage and the duration. We will also look at indicators of why spark duration is not sufficient. Keep in mind that problems in spark duration will usually result in misfire. Also keep in

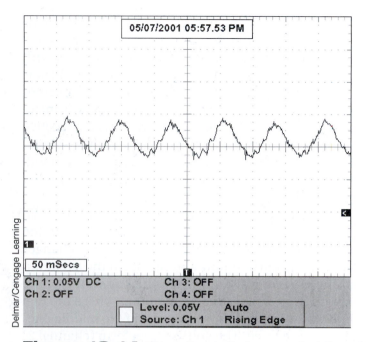

Figure 18-14 This pattern indicates normal relative compression.

mind that misfire may shut down the monitoring system. The DSO must have ignition adapters to allow it to capture and display the waveforms shown. Additionally, there are vehicles for which you will be unable to capture and use secondary ignition patterns. For these vehicles you may have to rely on primary voltage and current waveforms.

Figure 18-15 shows the firing voltage of a 6-cylinder engine.

The DSO has been set for ignition, and a synch probe has been place around spark plug wire #1 and will act as a trigger. The cylinders are labeled along the bottom in the firing order. Each division of voltage is 5 kV or 5,000 volts per division. Engines will generally have firing lines that run below 15 kV if there is mileage on the plugs. Brand new plugs will generally fire between 6 and 12 kV. Our highest firing voltage is 15 kV for cylinder #5, and our lowest one is around 12 kV for cylinder # 2. So far this pattern looks good. Remember, we are analyzing for OBD II problems that might prevent the monitors from running or cause misfire.

The next and most important piece to look at is burn time. *Burn time* is the time that the plug actually fires and is represented by the horizontal line following the firing line. Burn time must be greater than .8 mS (.0008 sec) or the cylinder may misfire. For this pattern we switch over to an expanded pattern for one cylinder or a raster pattern where patterns are stacked.

Figure 18-16 shows the burn time for one cylinder.

Cylinder #1 is highlighted so this pattern is for #1. The firing line is at the beginning of the 6th division and is followed by the burn time. With 1 mS (.001 sec) per division this plug appears to be firing for about 1.25 mS, which is above the required .8 mS.

Figure 18-17 shows all cylinders in raster display mode.

Now all cylinders can be compared for burn time. In this example, all cylinders appear to have about the same burn time of 1.25–1.5 mS. While we are examining these last two patterns, let's take a look at the primary on signal. The ignition module will turn on ignition coil primary current to charge the coil. Once the coil is charged, the ignition module will

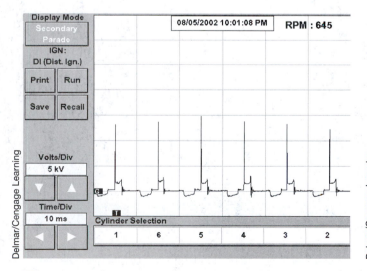

Figure 18-15 Most DSOs can display ignition firing voltages.

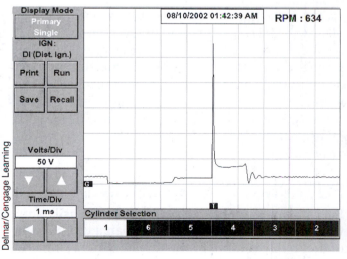

Figure 18-16 This pattern shows a 1.25 mS burn time, which is above the .8 mS minimum.

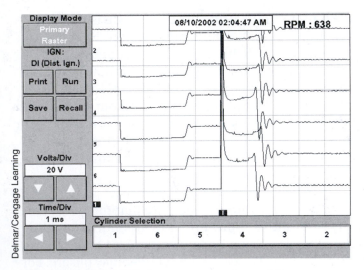

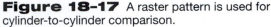

Figure 18-17 A raster pattern is used for cylinder-to-cylinder comparison.

turn off the current. This causes the magnetic field to collapse and generates the spark. Manufacturers have used a specification of 2.5 mS or greater primary on time for years. The signal shows on the DSO prior to the firing line and may be a downward turn of the trace that lasts all the way to the firing line or, as in our example, a downward turn of the trace that turns upward prior to the firing line. We are concerned with the time from downward turn until either upward turn or, if there is no upward turn, the firing line. In our examples in Figures 18-16 and 18-17 (both the same vehicle), the downward turn occurs in the beginning of the 2nd division and turns upward in the middle of the 4th division. So primary on time is 2.5 divisions long. Each division is equal to 1 mS, so our on time is 2.5 mS. A minimum of 2.5mS is common for GM vehicles with distributors, while a minimum of 3.5 mS is common for all other vehicles.

The most important point here is to note that a vehicle with OBD II problems in the ignition system may result in monitors not running or misfire DTCs. Our simple test of firing voltage, burn time, and primary on time is sufficient for "MIL on" testing or lack of run monitors, but would not be sufficient for no starts. No starts will require more extensive testing found in other texts like *Tech One: Automotive Electricity and Electronics*.

FUEL INJECTOR TESTING

Fuel system testing of the fuel injector is easily done with a DSO and offers the technician a unique view of the inner working of the injector, including how clean it is. To accomplish the testing, both a voltage

waveform and a current waveform will be used, as each can give valuable information. Figure 18-18 shows a waveform from a 6-cylinder vehicle, with multiport sequential fuel injection. Channel 1 (bottom waveform) is a current trace with the current probe placed around one of the wires to a single injector. Channel 2 is a voltage waveform with the positive test probe connected to the negative (driver side) of the injector and the black lead connected to ground. Each time the injector fires will result in the shown pattern.

Look at channel 1. As the fuel injector is turned on by the PCM, the current rises until the pintle in the injector opens. The movement of the pintle changes around the current flow and results in the slight dip of the current trace shown at the beginning of the 4th division. This is the actual point where the fuel injector opens and should be within the middle one-third of the trace. The voltage waveform (channel 2) shows the grounding of the fuel injector by the PCM to open it. This fuel injector waveform shows that the PCM wants the injector open for 5.25 mS. When the PCM removes the ground, there is an induced voltage just after the 6th division. This induced spike tapers down to battery voltage with a slight bump just after the 7th division. This is another pintle bump and results from the change in voltage when the fuel injector pintle closes. The time between the current pintle bump and the voltage pintle bump is the actual time that the fuel injector is open and

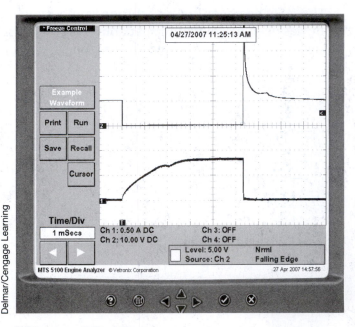

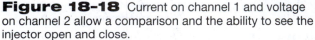

Figure 18-18 Current on channel 1 and voltage on channel 2 allow a comparison and the ability to see the injector open and close.

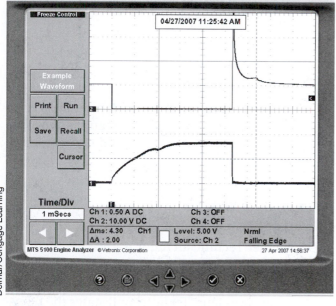

Figure 18-19 Cursors allow the measurement of injector on time.

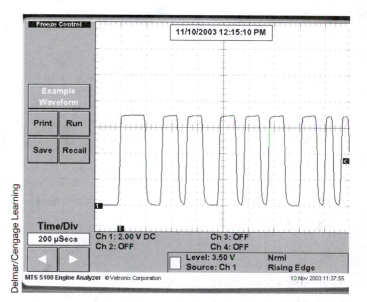

1	2	3	4	5	6	7	8
9	10	11	12	13	14	15	16

Figure 18-20 The data line communication's link for an OBD II system is a good place to scope serial data.
Delmar/Cengage Learning

Figure 18-21 The on/off signals from the DLC indicate good serial data.

should match the on time from the PCM. Using the DSO cursors will sometimes help you analyze the waveform and compare.

Figure 18–19 shows the same two traces but with the DSO cursors turned on.

One cursor is where the opening pintle bump is, and the other is where the closing pintle bump is. The delta mS shows at 4.30 mS. This is the actual time that the fuel injectors were open. In this example, the fuel injector is very dirty and does not flow fuel well, so the PCM increases the on time signal to 5.25 mS to get the required 4.30 mS worth of fuel. This is fuel trim in use. Cleaning this fuel injector will result in the two numbers getting closer. Very clean injectors will have a difference around .1 mS rather than the .95 mS in our example.

OBD II COMMUNICATION

Another good use for a DSO involves communication issues. Vehicles that are presented for the emissions test and will not communicate are either rejected or failed (based on different testing procedures in States). A vehicle that shows up in a shop without communication to a scanner needs to be fixed so that it will communicate. Figure 18–20 shows the DLC.

Pin 16 should have power, and pins 4 and 5 should be grounds. A DLC that is missing either power or ground will not communicate to most scanners and will not communicate in the emissions test lane.

The other pins are the communication pins with most OBD II vehicles communicating on 2, 7, 10, or 15. If the DSO is connected to pin 5 (ground), and the other lead is connected to a breakout box pin 2 or 7 or 10 or 15, the trace should show rapid on/off signals as in Figure 18–21.

This indicates communication. Remember, if a vehicle will not communicate in the test lane, the DLC is missing one of three requirements: power on pin 16, ground on pins 4 and 5, or a signal on either pins 2, 7, 10, or 15.

CONCLUSION

In this chapter, we looked at common uses of the digital storage oscilloscope. We analyzed signals from oxygen sensors and saw how they can be tested with propane. We then looked at using the DSO to calculate fuel pump speed followed by compression signals, ignition, and fuel injector waveforms. We ended with a look at the communication via the DLC and saw what a good signal would look like.

REVIEW QUESTIONS

1. Technician A states that a DSO reads voltage on the screen with eight divisions up and down. Technician B states that a DSO reads time over ten divisions from left to right. Who is correct?

   a. Technician A only

   b. Technician B only

   c. Both Technician A and B

   d. Neither Technician A nor B

2. A DSO is being set up to observe a 3X 18X signal off of a CKP. The screen shows eight patterns. What should be done?

   a. Nothing—the settings are correct.

   b. Increase the voltage per division.

   c. Decrease the time per division.

   d. Increase the time per division.

3. The correct settings for observing an O2S under test conditions is

   a. 200 mS and 500 mV/Div

   b. 200 mS and 200 mV/Div

   c. 500 mS and 500 mV/Div

   d. 500 mS and 200 mV/Div

4. A fuel pump shows eight commutator bumps in 14 mS. The approximate speed of the pump is

   a. 3200 RPM

   b. 4300 RPM

   c. 5100 RPM

   d. 6200 RPM

5. Two technicians are discussing fuel pump RPM. Technician A states that the pump should turn at 4500 RPM or greater. Technician B states that the pump should be capable of sufficient volume to allow the pressure regulator to function. Who is correct?

   a. Technician A only

   b. Technician B only

   c. Both Technician A and B

   d. Neither Technician A nor B

6. Compression testing is done with a DSO and a

   a. high-current probe

   b. low-current probe

   c. high-voltage probe

   d. low-voltage probe

7. Technician A states that spark plug burn time should exceed 2.5 mS. Technician B states that ignition primary on time should exceed .8 mS. Who is correct?

   a. Technician A only

   b. Technician B only

   c. Both Technician A and B

   d. Neither Technician A nor B

8. Technician A states that the bump on the current trace for a fuel injector is the point where the injector opens. Technician B states the bump on the voltage waveform for a fuel injector is the point where the injector closes. Who is correct?

   a. Technician A only

   b. Technician B only

   c. Both Technician A and B

   d. Neither Technician A nor B

9. Fuel injector on time is measured as 3.5 mS. Pintle bump open signal to pintle bump closed signal is measured as 4.6 mS. This indicates

   a. injectors are clean

   b. pressure regulator setting is low

   c. injectors need to be cleaned

   d. fuel pump volume is low

10. There is B+ at pin 16 of the DLC. There is ground on pin 4 of the DLC. Technician A states that this will cause a no start. Technician B states that this will cause no communication with most scan tools. Who is correct?

    a. Technician A only

    b. Technician B only

    c. Both Technician A and B

    d. Neither Technician A nor B

INDEX